JCER

Journal of Consciousness Exploration & Research

Volume 1 Issue 9

December 2010

Various Approaches to Consciousness & the Principle of Existence II

Chief Editor

Huping Hu, Ph.D., J.D.

Associate Editor

Maoxin Wu, M.D., Ph.D.

Editor-at-Large

Gregory M. Nixon, Ph.D.

ISSN: 2153-8212   Journal of Consciousness Exploration & Research   www.JCER.com
Published by QuantumDream, Inc.

Table of Contents

Journal of Consciousness Exploration & Research| December 2010 | Vol. 1 | Issue 9 | pp. 1-69
Conte, E., Todarello, O., Conte, S., Mendolicchio, L., Mendolicchio, L. & Federici, A.
Methods and Applications of Non-Linear Analysis in Neurology and Psycho-physiology

1

Article

Methods and Applications of Non-Linear Analysis in Neurology and Psycho-physiology

Elio Conte [(1,2)]*, Orlando Todarello[(3)], Sergio Conte[(2)],
Leonardo Mendolicchio[(4)], Antonio Federici[(1)]

[(1)]Department of Pharmacology and Human Physiology – TIRES – Center for Innovative Technologies for Signal
Detection and Processing, University of Bari- Italy;
[(2)]School of Advanced Int'l Studies for Applied Theoretical and Non Linear Methodologies of Physics, Bari, Italy;
[(3)] Department of Neurological and Psychiatric Sciences, University of Bari - Italy
[(4)] Department of Neurological and Psychiatric Sciences, University of Foggia - Italy

ABSTRACT

In the light of the results obtained during the last two decades in analysis of signals by time series, it has become evident that the tools of non linear dynamics have their elective role of application in biological, and, in particular, in neuro-physiological and psycho-physiological studies. The basic concept in non linear analysis of experimental time series is that one of recurrence whose conceptual counterpart is represented from variedness and variability that are the foundations of complexity in dynamic processes. Thus, the recurrence plots and the Recurrence Quantification Analysis (RQA) are discussed. It is shown that RQA represents the most general and correct methodology in investigation of experimental time series. By it we arrive to inspect the inner structure of the time series connected to the signals under investigation. Linked to RQA we prospect also the method CZF, recently introduced by us. It is able to account for a true estimation of variability of signals in time as well as in frequency domain. And, consequently, it may be used in conjunction with classical Fourier analysis, accounting however that it is inappropriate in analysis of non linear and non stationary experimental time series. The use of CZF method in fractal analysis is also considered in addition to standard index as Hurst exponent. A large field of possible applications in neurological as well as in psycho-physiological studies is given. Also, there are given examples of other and (possibly linked) applications as example the analysis of beat-to-beat fluctuations of human heartbeat intervals that is sovereign in psycho-physiological studies. We give applications on some different planes to evidence the particular sensitivity of such methods. We reach the objective to show that the previously exposed methods are also able to predict in advance the advent of ventricular tachycardia and/or of ventricular fibrillation. The RQA analysis gives good results. The CZF method gives the most excellent results showing that it is able to give very significant indexes of prediction. We also apply such methods in investigation of state anxiety, and proposing in detail a quantum like model of such phenomenological status of the mind.

Key Words: non-linear analysis, time series, neurology, psycho-physiology, RQA, CZF, anxiety, quantum-like, mind.

*Corresponding author: Elio Conte E-mail: elio.conte@fastwebnet.it

Journal of Consciousness Exploration & Research| December 2010 | Vol. 1 | Issue 9 | pp. 1-69   2
Conte, E., Todarello, O., Conte, S., Mendolicchio, L., Mendolicchio, L. & Federici, A.
Methods and Applications of Non-Linear Analysis in Neurology and Psycho-physiology

1. The Recurrences and the Variability of Signals in Nature

Only few systems in Nature exhibit linearity. The greatest whole of natural systems, especially those who pertain to biological matter, to physiological (neuro-physiological and psycho-physiological) and to psychological processes, possess a complexity that results in a great variedness and variability, linked to non linearity, to non stationarity, and to non predictability of their time dynamics. In the time domain, traditional methods were first used to describe the amplitude distribution of signals and later, methodologies used spectral analysis methods. However, they suffer of fundamental limits. They are applied assuming linearity and stationarity of signals that actually do not exist. The consequence is that such methods are unable to analyse in a proper way the irregularity present in most of signals. The results show that such irregularity is at the basis of the dynamics that we intend to explore. It reveals that complex behaviours of the system are very distant from previously accepted principles as it is the case, as example for biological signals, on the view of homeostatic equilibrium and of other similar mechanism of controls. The study of this very irregular behaviour requires the introduction of new basic principles. Therefore, nonlinear science is becoming an emerging methodological and theoretical framework that makes up what is called the science of the complexity, often called also chaos theory.

2. The Chaos Theory

The aim of non linear methodologies is a description of complexity and the exploration of the multidimensional interactions within and among components of given systems. An important concept here is that of chaotic behaviour. It will be defined chaotic if trajectories issuing from points of whatever degree of proximity in the space of phase, distance themselves from one another over time in an exponential way.

In detail, the basic critical principles may be reassumed as it follows:
   1) Non linear systems under certain conditions may exhibit chaotic behaviour;
   2) The behaviour of a chaotic system can change drastically in response to small changes in the system's initial conditions;
   3) A chaotic system is deterministic;
   4) In chaotic systems the output system is no more proportionate to system input.

Chaos may be identified in systems also excluding the requirement of determinism. The standard approach to classical dynamics assumes the Laplace point of view that the time evolution of a system is uniquely determined by it's initial conditions. Existence and uniqueness theorem of differential equations require that the equations of motion everywhere satisfy the Lipschitz condition. It has long been tacitly assumed that Nature is deterministic, and that correspondingly, the equations of motion describing physical systems are Lipschitz. However, there is no a priori reason to believe that Nature is unfailingly Lipschitzian. In very different conditions of interest, some systems exhibit physical solutions corresponding to equations of motion that violate the Lipschitz condition. The point is of particular interest. If a dynamical system is non-Lipschitz at a singular point, it is possible that several solutions will intersect at this point. This singularity is a common point among many trajectories, and the dynamics of the system, after the singular point

Journal of Consciousness Exploration & Research| December 2010 | Vol. 1 | Issue 9 | pp. 1-69   3
Conte, E., Todarello, O., Conte, S., Mendolicchio, L., Mendolicchio, L. & Federici, A.
Methods and Applications of Non-Linear Analysis in Neurology and Psycho-physiology

is intersected, is not in any way determined by the dynamics before. Hence the term non-deterministic dynamics takes place. For a non-deterministic system, it is entirely possible (if not likely) that as the various solutions move away from the singularity, they will evolve very differently, and tend to diverge. Several solutions coincide at the non-Lipschitz singularity, and therefore whenever a phase space trajectory comes near this point, any arbitrarily small perturbation may push the trajectory on to a completely different solution. As noise is intrinsic to any physical system, the time evolution of a non-deterministic dynamical system will consist of a series of transient trajectories, with a new one being chosen randomly whenever the solution (in the presence of noise) nears the non-Lipschitz point. We term such behaviour non-deterministic chaos. This approach to chaos theory was initiated by Zak, Zbilut and Webber [1] and rather recently we have given several examples, theoretical and experimental verifications on this important chaotic behaviour [2].

2.1 Embedding time series in phase space

The notion of phase space is well known in physics. Let us consider a system, determined by the set of its variables. Since they are known, those values specify the state of the system at any time. We may represent one set of those values as a point in a space, with coordinates corresponding to those variables. This construction of space is called phase space. The set of states of the system is represented by the set of points in the phase space. The question of interest is that we perform an analysis of the topological properties of phase space but, as a counterpart, we obtain insights into the dynamic nature of the system. In experimental conditions, especially in experimental clinical studies, we are unable to measure all the variables of the system. In this case we may be able to reconstruct equally a phase space from experimental data where only one of the present variables (characterizing the whole system) is actually measured. The phase space is realized by a set of independent coordinates. Generally speaking, the attractor is the phase space set generated by a dynamical system represented by a set of difference or differential equations. In the actual case, let us take a non linear dynamical system represented by three independent variables $X(t), Y(t), Z(t)$, functions of time. The phase space set is given by the values of the variables at each time. The point (x, y, z) in phase space gives the values of the three variables and thus the state of the system at each time. Usually, in physics, for example, we plot one of the variables and its derivatives,

$$X, \frac{dX}{dt}, \frac{d^2 X}{dt^2}, ...$$ (2.1)

on the three perpendicular axes (x, y, z). The result is that we have reconstructed the phase space using only one of the three time series using also the derivatives of $X(t)$. This is a licit step since $Y(t)$ and $Z(t)$ are coupled to $X(t)$ through non linear equations. Consider that in experiments we have a fixed time sampling, Δt (time series recorded at equal time intervals), and the time series is given in the following manner

$$X(0), X(\Delta t), X(2\Delta t), X(3\Delta t), ..., X(n\Delta t)$$ (2.2)

We could also differentiate such values determining $dX/dt, d^2 X/dt^2, ...$but such a procedure

Journal of Consciousness Exploration & Research| December 2010 | Vol. 1 | Issue 9 | pp. 1-69
Conte, E., Todarello, O., Conte, S., Mendolicchio, L., Mendolicchio, L. & Federici, A.
Methods and Applications of Non-Linear Analysis in Neurology and Psycho-physiology

4

is unprofitable. In fact, also if our time series data should contain only very small errors in measurements, they should become larger errors during such operations. We may follow another procedure. We introduce a time lag $\tau = m\Delta t$ and consider each point in phase space, given by the following vector expression

$$[X(t), X(t+\tau), X(t+2\tau), X(t+3\tau), \ldots \ldots X(t+(N-1)\tau)] = \overline{X}_N(t) \tag{2.3}$$

where N is the selected dimension of the phase space. Note that assuming such a procedure in phase space reconstruction we do not lose generality since, as it is easy to show, the coordinates of the phase space reconstructed in this manner, using time delays, are linear combinations of the derivatives.

This procedure of reconstruction of phase space starting with the given time series is called *embedding*. This is the method presently used for reconstruction of phase space of experimentally sampled time series. Takens in 1981 [4] showed that this embedding method, based on time lags, is certainly valid under some suitable conditions. The first requirement is that the considered time series must be twice differentiable. If this requirement is not satisfied, and it happens often, when the considered time series is a fractal, the fractal dimension, calculated by the embedding method, may also not be equal to the true fractal dimension of the phase space set. Still, the other statement relating Takens theorem, requires that in a realistic reconstruction of phase space, say of dimension D, we must embed in a space of dimension $(2D+1)$ in order to express enough dimensions. This is to avoid the possibility that the $N-$ dimensional orbits intersect themselves in a false manner.

2.2 The Determination of Time Lag τ

Some different procedures may be followed to determine the time lag of the given time series in the embedding method. There are cases in which the appropriate choice of the time lag is rather simple. In fact, it may be seen from the basic features of the system under consideration. It is rather simple to evaluate the proper time lag if we are investigating physiological processes exhibiting with evidence their natural time scale. In other cases the estimation of the time lag may be not be so simple since we do not have a direct indication of the appropriate time lag. Let us consider, for example, the case of investigation of a physiological process involving electroencephalographic studies.

Experience in methodological analysis of time series often helps to solve such problems. The problem must be solved with particular care. The proper choice of the time lag is of fundamental importance because in chaotic signals the relation between the dimension of an embedding space and real phase space is strongly linked to the length chosen for a time lag. A too large selected time lag will determine unwished noise in embedding and so the observation of the chaotic attractor will be strongly compromised. The use of a too small lag may result in strong correlations among the components of the signal (2.3), and the local geometry of embedding results much like as a line (i.e. dimension equal to 1), and damaging image reconstruction of the chaotic attractor.

Journal of Consciousness Exploration & Research| December 2010 | Vol. 1 | Issue 9 | pp. 1-69
Conte, E., Todarello, O., Conte, S., Mendolicchio, L., Mendolicchio, L. & Federici, A.
Methods and Applications of Non-Linear Analysis in Neurology and Psycho-physiology

5

As a methodological praxis, it is useful to study the autocorrelation function of the given time series. Given the time series $X(n)$, $n=1, 2, ...N$, the autocorrelation function, $Au(\tau)$, at lag τ is defined as:

$$Au(\tau) = \frac{1}{N-\tau} \sum_{n=1}^{N-\tau} X(n)X(n+\tau) \tag{2.4}$$

Values of time series correlate with themselves and the correlation diminishes as the time lag between two points increases. Correlation decreases with time. The time lag is selected as the autocorrelation function reaches its first zero. Often another useful criteria is to take the time lag as the autocorrelation function decreases to $1/e = 0.37$.

In addition to use of the autocorrelation function, one can employ the mutual information content, $MI(\tau)$. Mean mutual information is given in the following manner [5]

$$MI(\tau) = \sum_{X(i),X(i+\tau)} P(X(i), X(i+\tau)) \log_2 \frac{P(X(i), X(i+\tau))}{P(X(i))P(X(i+\tau))} \tag{2.5}$$

The time at which the first local minimum of mutual information content is reached, may represent a good choice for time lag. Both $Au(\tau)$ and $MI(\tau)$ must be used, selecting the time lag provided by $MI(\tau)$ if $Au(\tau)$ and $MI(\tau)$ predict different results. This is preferable since $MI(\tau)$ also accounts for non linear contributions in a time series.

2.3 Embedding Theorem and False Nearest Neighbors

As previously outlined, according to the embedding theorem (see Takens theorem for details), the choice of dimension N of reconstructed phase space should require a priori knowledge of the dimension d_F of the original attractor with $N > 2d_F$. This is decisively unrealistic for time series of experimental data. Selecting N in absence of a given criterion, it may result in too small a choice as compared to the d_F of the original attractor. It is possible to employ what is called the criterion of false nearest neighbors (FNN) in reconstructed phase space [6]. A point of data sets is said to be a FNN when it comprises the local nearest neighbors not actually but only because the orbit is constructed in a too small an embedded space determining its self-crossing. This difficulty may be overcome by adding sufficient coordinates to the embedding space. The criterion to use is to increase N in a step manner until the number of the FNN goes substantially to zero. Usually, a threshold of about 5% may be acceptable. Le us calculate the distance between two points in a selected embedding dimension of N, obtaining the value $D_N(i)$. In the $(N+1)$ embedding dimension, we will have $D_{N+1}(i)$. Such values satisfy the following relation

$$\sqrt{\frac{D_{N+1}^2(i) - D_N^2(i)}{D_N^2(i)}} = \frac{\left| X_{i+N\tau} - X_{i+N\tau}^{NN} \right|^2}{D_N(i)} \tag{2.6}$$

where NN indicates that we consider a point selected conventionally near a given point. A fixed

Journal of Consciousness Exploration & Research| December 2010 | Vol. 1 | Issue 9 | pp. 1-69   6
Conte, E., Todarello, O., Conte, S., Mendolicchio, L., Mendolicchio, L. & Federici, A.
Methods and Applications of Non-Linear Analysis in Neurology and Psycho-physiology

threshold value is used and step by step it is verified if the (2.6) exceeds or not the prefixed threshold value.

3. Fractality and Non Linearity of Experimental Time Series

3.1 Fractality and Deterministic chaos of Time Series

The use of non linear methods presumes that the signal under study is represented by an experimental time series relating a non linear system. Sometimes it possesses some deterministic features that may be also chaotic and must be investigated by the methodology discussed in the previous sections.

Fractality refers to the features of a given stochastic time series. It shows temporal self-similarity. A time series is said self-similar if its amplitude distribution remains unchanged by a constant factor even when the sampling rate is changed. In the time domain one observes similar patterns at different time scales. In the frequency domain the basic feature of a fractal time series is its power law spectrum in the proper logarithmic scale. Fractals and chaos have many common points. When the phase space set is fractal, the system that generated the time series is chaotic. Chaotic systems can be arranged that generate a phase space set of a given fractal form. However, the systems and the processes studied by fractals and chaos are essentially different. Fractals must be considered processes in which a small section resembles the whole. The point in fractal analysis is to determine if the given experimental time series contains self-similar features. Deterministic chaos means that the output of a non linear deterministic system is so complex that in some manner mimes random behaviour. The point in deterministic chaos analysis is to investigate the given experimental time series that arises from a deterministic process and to understand in some manner the mathematical features of such a process. Regarding a chaotic time series, this means that the corresponding system has sensitivity to initial conditions. When we speak about strange attractors this means that the attractor is fractal [for details see 4]. It is very important to account for such properties since there are also chaotic systems that are not strange in the sense that they are exponentially sensitive to initial conditions but do not have a fractal attractor. Still we have non chaotic systems that are strange in the sense that they are not sensitive to initial conditions but they have a fractal attractor. In conclusion, we must be careful in considering fractals and non linear approaches since they are very different from each other. Often, instead, we are induced to erroneously mix different things with serious mistakes.

The geometry of the attractors is frequently examined by calculating the so called correlation dimension [7]. The self-similar property of the attractor is estimated by the scaling behaviour of the correlation integral

$$C_N(r) = \frac{1}{n^2} \sum_{i \neq j} \vartheta \left(r - \left\| \overline{X}_N(i) - \overline{X}_N(j) \right\| \right)$$

where $\vartheta(\cdot)$ is 1 for positive arguments and 0 for negative arguments. For a fixed a sphere of

Journal of Consciousness Exploration & Research| December 2010 | Vol. 1 | Issue 9 | pp. 1-69   7
Conte, E., Todarello, O., Conte, S., Mendolicchio, L., Mendolicchio, L. & Federici, A.
Methods and Applications of Non-Linear Analysis in Neurology and Psycho-physiology

radius r, in the reconstructed phase space $C_N(r)$ gives the normalized number of points in it. For stochastic signals the correlation integral, calculated in the N – dimensional space, scales as

$$C_N(r) \approx r^N$$

For bounded signals there is a finite scaling exponent so that

$$C_N(r) \approx r^d \text{ with } d < N .$$

The correlation dimension, usually indicated by D_2, is calculated as the slope of the linear behaviour of $\log r$ vs. $\log C_N(r)$. The value 1.0 is obtained in the case of a limit cycle, 2.0 instead is calculated in the case of a torus. A calculated non- integer value instead indicates that the phase space has a fractal geometry. However, in analysis of experimental time series the calculation of the correlation dimension does not offer results sensitive enough to conclude that for a non-integer, a fractal dimension that could be generated by a deterministic chaotic system. Stochastic signals may mimic chaotic data and furthermore, time series of stationary data are always required. This last requirement is rarely obtained by experimental time series, especially those of biological or physiological interest.

3.2 Estimation of Lyapunov Exponents

As previously mentioned, chaotic systems show a dynamics where phase space trajectories with nearly identical initial states will, however, separate from each other at an exponentially increasing rate. This is usually called the sensitive dependence on initial conditions in chaotic deterministic systems. The spectrum of the Lyapunov exponents captures this basic feature of the dynamics of these systems. We may consider the two nearest neighboring points in phase space at time 0 and at time t. Let us consider also a direction i-th in space. Let $\|\delta x_i(0)\|$ be the distance at time 0 and $\|\delta x_i(t)\|$ the distance at time t. The Lyapunov exponent, λ_i (direction i-th), will be calculated such that [8]

$$\frac{\|\delta x_i(t)\|}{\|\delta x_i(0)\|} = e^{\lambda_i t} \qquad \text{for } t \to \infty$$

that is equivalent to

$$\lambda_i = \lim_{t \to \infty} \frac{1}{t} Ln \frac{\|\delta x_i(t)\|}{\|\delta x_i(0)\|}$$

It is possible to reconstruct the Lyapunov spectrum accounting for all the directions in phase space. Chaotic systems are characterized by having at least one positive Lyapunov exponent while their sum generally must be negative. Given there is a whole spectrum of Lyapunov

Journal of Consciousness Exploration & Research| December 2010 | Vol. 1 | Issue 9 | pp. 1-69
Conte, E., Todarello, O., Conte, S., Mendolicchio, L., Mendolicchio, L. & Federici, A.
Methods and Applications of Non-Linear Analysis in Neurology and Psycho-physiology

8

exponents, the number of them is equal to the number of dimensions of the phase space. If the system is conservative (i.e. there is no dissipation), a volume element of the phase space will stay the same along a trajectory. Thus the sum of all Lyapunov exponents must be zero. If the system is dissipative, the sum of Lyapunov exponents is negative.

The Lyapunov spectrum can be used also to give an estimate of the rate of entropy production and of the fractal dimension of the considered dynamical system. In particular from the knowledge of the Lyapunov spectrum it is possible to obtain the so-called Kaplan-Yorke dimension D_{KY}, that is defined as follows:

$$D_{KY} = k + \sum_{i=1}^{k} \frac{\lambda_i}{|\lambda_{k+1}|}$$

where k is the maximum integer such that the sum of the k largest exponents is still non-negative. D_{KY} represents an upper bound for the information dimension of the system. Moreover, the sum of all the positive Lyapunov exponents gives an estimate of the Kolmogorov-Sinai entropy accordingly to Pesin's theorem [9]

In conclusion, the Lyapunov exponent is a measure of the rate at which nearby trajectories in phase space diverge. Chaotic orbits show at least one positive Lyapunov exponent. Instead periodic orbits all give negative Lyapunov exponents. It is of interest also the analysis of a Lyapunov exponent equal to zero. It says that we are near a bifurcation.

There is still another feature to outline. It is common to avoid to calculating the whole Lyapunov spectrum, estimating instead only the most positive one, usually refered to as the largest one. A positive value is normally taken as indication that the system is chaotic. The inverse of the largest Lyapunov exponent is sometimes referred to in the literature as Lyapunov time, and defines the characteristic folding time. For chaotic orbits it is finite, whereas for regular orbits it will be infinite. Finally, to quantify predictability of the system, the rate of divergence of the trajectories in phase space must be evaluated by Lyapunov exponents and Kolmogorov-Sinai entropy.

Under the perspective of the analysis one must account for the calculation of Lyapunov exponents from limited experimental data of time series. Various methods have been proposed [10]. Generally speaking, however, these methods may be sensitive to variations in parameters, e.g., number of data points, embedding dimension, reconstructed time delay, and are usually reliable with care.

3.3 The Method of Surrogate Data in Time Series

At this stage of the present exposition, the reader will have realized that the most unfavourable snare in the investigation of experimental time series, possibly chaotic, is that the methods we have at our availability, are inclined to give similar results in the case of deterministic chaotic dynamics and stochastic noise so that distinguishing deterministic chaos from noise becomes an important problem. Starting with a given experimental time series, stochastic surrogate data may be generated having the same power spectra as the original one, but having random phase relationship among the Fourier components. If any numerical procedure in studying deterministic-chaotic dynamics will produce the same results for surrogate data as well as for the

Journal of Consciousness Exploration & Research| December 2010 | Vol. 1 | Issue 9 | pp. 1-69   9
Conte, E., Todarello, O., Conte, S., Mendolicchio, L., Mendolicchio, L. & Federici, A.
Methods and Applications of Non-Linear Analysis in Neurology and Psycho-physiology

original ones within a prefixed criterion, we will not reject the null hypothesis that the analyzed dynamics is determiend by a linear stochastic model rather than to be represneted by deterministic chaos. Often the method of the shuffled data is used. Data of the original time series are shuffled, and this operation preserves the probability distribution but produces generally a very different power spectrum and correlation function.

3.4 Fractional Brownian Analysis in Time Seires

It is well knwn that the study of stochastic processes with power-law spectra started with the celebrated paper on fractional Brownian motion (fBm) by Mandelbrot and Van Ness in 1968 [11]. Fixing the initial conditions, fBm is defined by the following equation

$$X(ht) \overset{d}{=} h^H X(t) \tag{3.1}$$

Given a self-similar fractal time series, (3.1) establishes that the distribution remains unchanged by the factor h^H even after the time scale is changed. $(\overset{d}{=})$ states that the statistical distribution function remains unchanged. H is called Hurst exponent, varying as $0 < H < 1$, and it characterizes the general power – law scaling. For an additive process of Gaussian white noise, we have $H = 0.5$. H values greater than 0.5 indicate persistence in time series. This is to say that a past trend persists into the future(long-range correlation). Instead, H values less than 0.5 indicate antipersistence and this is to say that past trends tend to reverse in the future. The fBm also exhibits power-law behaviour in the Fourier spectrum. There is a linear relationship between the log of spectral power vs. log of frequency. The inverse of the slope in the log-log plot is called the spectral exponent β $(1/f^\beta$ behaviour), and it is related to H by the following relationship

$$H = \frac{\beta - 1}{2}$$

4. Recurrence Quantification Analysis and the CZF Method

4.1 Introduction

Let us take up some of the concepts exposed in the previous sections. It was outlined that the most important concept in studies of nonlinear processes by time series is that one of recurrence. A recurrence plot is the visualization of a square recurrence matrix of distance elements within a cutoff limit. We outlined also the importance of Takens theorem relating higher dimensional reconstruction of signals by the method of time delay. It is important to reaffirm here that the topological features of a higher dimensional system consisting of multiple coupled variables may be reconstructed from a single measured variable. We measure only one of these variables, and correspondingly we obtain important information on the whole system underlying the dynamics. The reconstruction happens in the phase space. Let us discuss an example previously introduced in [3] to illustrate the importance of the approach.

Journal of Consciousness Exploration & Research| December 2010 | Vol. 1 | Issue 9 | pp. 1-69
Conte, E., Todarello, O., Conte, S., Mendolicchio, L., Mendolicchio, L. & Federici, A.
Methods and Applications of Non-Linear Analysis in Neurology and Psycho-physiology

10

Let us take a single lead of the ECG recorded signal. We have in this manner an ECG signal in its one dimensional representation of voltage as a function of time. A digitised time series is obtained. Actually ECG derives from summed cardiac potentials that act simultaneously under the frontal, the saggital and the horizontal orthogonal planes, and thus along three dimensions. In order to have an accurate representation of the ECG signal, we need to simultaneously record voltages in time in these three orthogonal planes. However, if we perform a reconstruction plotting 1-dimensional data again itself and twice delayed, that is to say, delayed by τ and 2τ, on a three axis plot, the signal is represented as the reconstructed 3-dimensional space. Topologically, these loops are the same thing as the simultaneous plotting of three orthogonal recorded ECG leads. In the previous sections, we outlined that in order to realize such a methodology we need to estimate properly the time delay and the embedding dimension.

In analysis, recurrence is the most important concept. Of course, variedness and variability relate the complexity of a given dynamics. In recurrence analysis one must define some parameters that are the range, the norm, the rescaling and, finally, the radius, and the line. The range defines a window on the dynamics under investigation, selecting the starting point and the ending point in the time series to be analysed. For the norm, one has to distinguish the minimum norm, the maximum and the Euclidean norms. The norm function geometrically defines the size and the shape of the neighborhood surrounding each reference point. The Euclidean norm defines the Euclidean distance between paired points in phase space. The rescaling relates the fact that the distance matrix can be rescaled by dividing each element in the distance matrix by either the mean distance or maximum distance of the whole matrix. Finally, the radius is expressed in units relative to the elements in the distance matrix, whether or not these elements have been rescaled. The line parameter is decisive when we have to extract quantitative features from recurrence plots We have a length of a recurrence feature and a prefixed line parameter so that such features may be rejected in quantitative analysis if it results are shorter than selected line parameter.

4.2 The Recurrence Quantification Analysis

Recurrence analysis was first introduced by Eckmann, Kamphorst and Ruelle in 1987 [12], A recurrence quantification analysis, indicated by RQA, was subsequently introduced by Zbilut and Webber [13] and further enriched by the introduction of other variables by Marwan [14]. An exceptional element of value of RQA is that this method has no restrictions in its applications: as we will explain later, for example it may be applied also to non stationary time series.

The first recurrence variable is the % Recurrence (%REC). %REC quantifies the percentage of recurrent points falling within the specified radius. Out of any doubt we may define it the most important variable in analysis of time series. The second recurrence variable is the % Determinism (%DET). %DET measures the proportion of recurrent points forming diagonal line structures. Diagonal line segments must have a minimum length in relation to above line parameter. Repeating or deterministic patterns are characterized by this variable. Periodic signals will give long diagonal lines. Instead chaotic signals will give very short diagonal lines. Stochastic signals will not determine diagonal lines unless a very high value of the radius will be selected.

Journal of Consciousness Exploration & Research| December 2010 | Vol. 1 | Issue 9 | pp. 1-69   11
Conte, E., Todarello, O., Conte, S., Mendolicchio, L., Mendolicchio, L. & Federici, A.
Methods and Applications of Non-Linear Analysis in Neurology and Psycho-physiology

The third recurrence variable is the MaxLine (LMAX). It is the length of the longest diagonal line segment in the plot excluding obviously the main diagonal line of identity. This is a variable of interest since it inversely scales with the most positive Lyapunov exponent previously discussed. Therefore, the shorter the maxline results, the more chaotic the signal is. In addition, RQA may be performed by epochs, so that LMAX enables evaluation of Lyapunov exponent locally.

The other important recurrence variable is entropy (ENT). It relates Shannon information entropy of all the diagonal line lengths distributed over integer bins in a histogram. ENT may be considered a measure of the signal complexity and is given in bits/bin. For simple periodic systems with all diagonal lines of equal length and the entropy is expected to go to zero.

Another decisive variable in RQA is the trend (TND). All the above methods discussed in the previous sections hold for stationary time series. This is a condition rarely met in analysis of experimental time series and especially in the field of biological signals. RQA may be applied for any kind of experimental time series including non stationary time series. This is one of the reasons to appreciate the RQA method. The trend (TND) still quantifies the degree of non stationarity of the time series under investigation. If recurrent points are homogeneously distributed across the recurrence plot, TND values will approach zero. If they are heterogeneously distributed across the recurrence plot, TND values will result different from zero.

The sixth important variable in RQA, introduced by Marwan [14] is %Laminarity (%LAM). %LAM measures the percentage of recurrent points in vertical line structures rather than diagonal line structures. Finally, the Trapping Time (TT) measures the average length of vertical line structures. Square areas (really a combination of vertical and diagonal lines) indicate laminar (singular) areas, possibly intermittency, suggesting transitional regimes, chaos-ordered, chaos-chaos transitions.

In conclusion, RQA may be considered at the moment the most powerful method for analysis of any kind of time series without limitations of any kind. The confirmation is in the large and growing interest in literature for such a methodology over the last decade. Several fields have been explored by RQA from general chaos science to proper fields of application as clinical electro-physiology [see as example 15], molecular dynamics, psychology and mind pathologies [see for example 16], finance, just to list only some of the several fields impacted by this non linear methodology of analysis.

4.3 Further Advances in Analysis of Variability in Time Series: the CZF method

As previously indicated, complexity of natural processes relates the variedness and the variability of the experimentally measured signals in the form of time series. The CZF method relates this feature, and it derives from the surname (Conte, Zbilut, Federici) of the authors who introduced it.

Let us recall an old notion. The presence of an harmonic component in a given time series is

Journal of Consciousness Exploration & Research| December 2010 | Vol. 1 | Issue 9 | pp. 1-69   12
Conte, E., Todarello, O., Conte, S., Mendolicchio, L., Mendolicchio, L. & Federici, A.
Methods and Applications of Non-Linear Analysis in Neurology and Psycho-physiology

revealed by its power spectrum $P(v)$ given by the squared norm of the Fourier transform of the given time series $X(t)$ as

$$P(v) = \left\| \int_0^\infty e^{iv t} X(t) dt \right\|^2 \tag{4.1}$$

and evidences sharp peaks.

FFT (Fast Fourier Transform), in its discrete version, is currently applied in analysis of non linear time series. All we know that, because of its simplicity, Fourier analysis has dominated and still dominates the data analysis efforts. This happens ignoring the fact that FFT is valid under extremely general conditions but essentially under the respect of some crucial restrictions that often result largely violated, especially in the field of the electrophysiological signals. Three stringent conditions must be observed:
1) the system under investigation must be linear.
2) The data of the time series under investigation must be strictly periodic and stationary.
3) All the data of the time series under investigation must be sampled at equally spaced time intervals.

The consequences of such improper use of the FFT are significant. In particular, the presence of non linearity and of non stationarity give little sense to the results that are obtained. Consequently we will discuss now a non linear method, the CZF. It was previously introduced by us in literature [17], and it presents, conceptual links with RQA.

Let us start with Hurst analysis [18] that brings light on some statistical properties of time series $X(t)$ that scale with an observed period of observation T and a time resolution μ. As previously shown, scaling results characterized by an exponent H that relates the long-term statistical dependence of the signal. In substance, one may generalize such Hurst approach, expressing the scaling behaviour of statistically significant properties of the signal. Indicating by E the mean values, we have to analyze the q-order moments of the distribution of the increments

$$K_q(\tau) = \frac{E\left(|X(t+\tau) - X(t)|^q \right)}{E\left(|X(t)|^q \right)} \tag{4.2}$$

The (4.2) represents the statistical time evolution of the given stochastic variable $X(t)$.
For q=2, we may re-write the (4.2) in the following manner

$$\gamma(h) = \frac{1}{2n(h)} \sum_{i=1}^{n(h)} \left[X(u_{i+h}) - X(u_i) \right]^2 \tag{4.3}$$

that estimates the variogram of the given time series. Here, $n(h)$ is the number of pairs at lag distance h while $X(u_i)$ and $X(u_{i+h})$ are time sampled series values at times t and $t+h$,

Journal of Consciousness Exploration & Research| December 2010 | Vol. 1 | Issue 9 | pp. 1-69
Conte, E., Todarello, O., Conte, S., Mendolicchio, L., Mendolicchio, L. & Federici, A.
Methods and Applications of Non-Linear Analysis in Neurology and Psycho-physiology

13

$t = u_1, u_2, \ldots ; \quad h = 1, 2, 3, \ldots$. In substance, the variogram is a statistical measure expressed in the form:

$$\gamma(h) = \frac{1}{2} Var\left[X(u+h) - X(u)\right] \tag{4.4}$$

The variogram here introduced represents the a valuable measure of complexity in a given non linear time series and at the same time its elaboration enables us to overcome the difficulties previously mentioned for use of the FFT in non stationary and non linear time series. The concept of variability is sovereign in this case. Let us take an example to illustrate its relevance. Let us admit we have a time series given only by six terms:

$$X_1, X_2, X_3, X_4, X_5, X_6 . \tag{4.5}$$

The first time we select time lag $h = 1$, and using the (4.3) we calculate variability of this signal at this time scale, obtaining:

$$(X_1 - X_2)^2 + (X_2 - X_3)^2 + (X_3 - X_4)^2 + (X_4 - X_5)^2 + (X_5 - X_6)^2 \tag{4.6}$$

This is the variability of the signal at time scale $h = 1$ and, in accord with the (4.3), we indicate it by $\gamma_1(h) \equiv \gamma_1(1)$.

Note some important features:
The differences $(X_i - X_{i+1})^2$ in the (4.6) will account directly for the fluctuations (and thus of the total variability) that intervene in X_{i+1} with respect to X_i. It will be due to the particular features of the dynamics under investigation. Let us consider for example the case of (4.5) representing the beat-to-beat fluctuations of human heartbeat intervals. The (4.6) will represent total variability in time lag $h = 1$ due to the regulative activity exercised by sympathetic, vagal, and VLF activities in the time lag considered. Still, the count of such variability will happen for all the points of the given time series and thus it will account for the total variability at the fixed time scale of resolution for the whole considered R–R process.

Finally, if $\gamma_1(1)$ will assume a value going to zero, we will conclude that at such time scale (time lag delay h = 1) the variability of the signal in this time lag is very modest. Otherwise, if $\gamma_1(1)$ is different from zero in a consistent way, we will conclude that it gives great variability, attributed to the presence of a relevant activity of control. In the same way we will proceed considering for example (4.5) to represent an EEG signal recorded at some electrode at a given sampling frequency. In this case, (4.6) represents the total variability in cerebral activity at the selected electrode and at the time resolution of $h = 1$. After to having computed the total variability of signals at this time resolution $h = 1$, we will continue our calculation evaluating this time the total variability of the signal at the time resolution $h = 2$, and thus calculating $\gamma_2(2)$. In a similar way we will proceed calculating total variability at the time scale resolution corresponding to $h = 3$ and so on, completing the analysis of variability at each time scale. In conclusion we will

Journal of Consciousness Exploration & Research| December 2010 | Vol. 1 | Issue 9 | pp. 1-69

14

Conte, E., Todarello, O., Conte, S., Mendolicchio, L., Mendolicchio, L. & Federici, A.

Methods and Applications of Non-Linear Analysis in Neurology and Psycho-physiology

calculate the final variability of the given signal step by step at different time scales. The result will be a diagram in a plot in which in axis of the ordinate we will have the values of variability (in its corresponding unity of measurement) while in the axis of the abscissa we will have the corresponding value of h, that is to say of the corresponding time resolution.

Note that, in order to calculate the final value of the total variability we may decide (at time lag $h = 1$ but so also at the following steps) to divide (4.6) by the number of pairs employed in the calculation. In this manner we will obtain the mean value of variability at such time scale.

To complete our exposition on the CZF method we must still outline that, in calculating $\gamma_i(h)$ we may also use the embedding procedure for reconstruction in phase space and thus performing in this case a more elaborate and significant exploration of the time series under investigation.

From a methodological view point we may still outline that by the CZF method we may perform also fractal analysis of the given time series. In fact we may use the Fractal Variance Function, $\gamma(h)$, and the Generalized Fractal Dimension, $D_{\dim}$, by the following equation

$$\gamma(h) = Ch^{D_{\dim}} \tag{4.7}$$

and finally estimating the Marginal Density Function for self-affine distributions, given by the following equation [19]:

$$P(h) = ak^{-a}h^{a-1} \tag{4.8}$$

This last consideration completes our exposition on CZF method. It remains to be explained the manner in which the CZF method overcomes the difficulties previously noted in the case of FFT and thus the manner in which it must be applied to perform an analysis of variability in the frequency domain. To illustrate such a methodology we will use two basic examples: the first is the case of HRV, that is the analysis of heart rate variability by using time series of R-R intervals from the ECG. The second example will relate the analysis of variability of brain waves in EEG in the frequency domain.

4.4 An Example of Application of CZF method in HRV analysis of R-R time series from ECG

It is well known that R-R time series relate the beat-to-beat time fluctuations of human heartbeat intervals and R-R values are largely controlled by various physiological and psychological factors and, in particular, by the balance between sympathetic and parasympathetic nervous system activity imposed upon the spontaneous discharge frequency of the sinoatrial node.R-R analysis is largely used in psycho-physiological studies. We quote only two papers to outline the importance of such field. The first is an analysis of cardiac signature of emotionality as quoted in ref.24. The second is an analysis of heart period variability and depressive symptoms:gender differences as quoted in ref.25.

Fluctuations in time in R-R result in what we call the variability of the R-R signal and, using the FFT, in the frequency domain three bands are identified. The first, the VLF, is usually considered

Journal of Consciousness Exploration & Research| December 2010 | Vol. 1 | Issue 9 | pp. 1-69
Conte, E., Todarello, O., Conte, S., Mendolicchio, L., Mendolicchio, L. & Federici, A.
Methods and Applications of Non-Linear Analysis in Neurology and Psycho-physiology

15

to range from 0 to 0.04 Hz and related to humoral regulation of the sinus pacemaker cell activity and to other contributing factors; the second, the LF, ranging from 0.04 to 0.15 Hz, and the HF, ranging from 0.15 to 0.4 Hz are roughly correlated to autonomic sympathetic and vagal activities, respectively.

To perform analysis of variability by CZF in the frequency domain we calculate the mean value, $E(R-R)$, in msec. Consequently we will estimate an equivalent frequency

$$f_{equivalent} = \frac{1}{E(R-R)}$$

Finally, we realize the final diagram having on the ordinate the values of the variability as calculated by (4.3) and on the abscissa, in correspondence with each lag , h, we will assign instead the value $hf_{equivalent}$ with $h = 1,2,3,....$

We will now apply the CZF method to the case of the beat-to-beat fluctuations of human hearthbeat intervals in the cases of normal subjects and subjects with pathologies. We will give the CZF results after having performed the analysis of the given R-R time series using also the previously explained other methodologies. We selected four groups of five subjects. Data were taken from Physionet [20].

Let us delineate some features of the experimental data. The first two groups, Y_i and O_i (i=1, 2, 3, 4, 5), are young and old subjects, respectively. Young subjects were (21 – 34) years old and old subjects were (68 – 85) years old. Men and women were included in the two groups All were rigorously-screened and found to be healthy subjects. ECG recording was performed for 120 minutes of continuous supine resting. The continuous ECG, respiration, and (where available) blood pressure signals were digitized at 250 Hz. Each heartbeat was annotated using an automated arrhythmia detection algorithm, and each beat annotation was verified by visual inspection. We selected pieces of 1024 R-R data points corresponding to a time interval of about thirteen minutes. For the other two groups, Vt_i and Vf_i, instead pieces of 1024 data points of R-R time intervals were chosen immediately before the advent of an episode of ventricular tachycardia (V_t) and ventricular fibrillation (V_f).

The first step was to apply the embedding procedure for phase space reconstruction of the given R-R signals. As previously explained, we calculated first the Autocorrelation Function (Au), then the Mean Mutual Information (MI) to select a proper time delay τ. When the results predicted by Au and MI were different, we opted for the time delay as predicted from MI. After this step, we proceeded to the final phase space reconstruction by using the criterion of False Nearest Neighbors (FNN) fixing a threshold value. In order to give some indication, in Figures 1a, 1b, 1c, we give the AutoCorrelation function (Au), MI, and FNN results for some subject of group Yi, In Figures 2a, 2b, 2c, the corresponding results for a subject of group O_i, and, finally, in Figures 3a, 3b, 3c, and Figures 4a, 4b, 4c, those for a subject in group V_{ti} and a subject in group V_{fi}, respectively. All the results are given in Table 1. In spite of different values obtained for Au, it may be seen that rather constant values of time delays were obtained by using MI. They ranged between 1 and 3 for young and old healthy subjects with an embedding dimension that resulted in

Journal of Consciousness Exploration & Research| December 2010 | Vol. 1 | Issue 9 | pp. 1-69
Conte, E., Todarello, O., Conte, S., Mendolicchio, L., Mendolicchio, L. & Federici, A.
Methods and Applications of Non-Linear Analysis in Neurology and Psycho-physiology

16

being constantly equal to 5 for young subjects, and constantly equal to 4 for old subjects. The subjects in V_t gave time delays ranging between 2 and 4 but this time the embedding dimension resulted in varying from 2 to 7. V_f subjects gave time delays between 2 and 4 but the embedding dimension varied from 2 to 8.

Phase space reconstruction resulted in rather homogeneous results in the group of normal subjects, the O_i group, and the Y_i group, with differences in embedding dimension in old subjects (embedding dimension equal to 4) with respect to young subjects (embedding dimension equal to 5). Instead, marked differences arose in the groups V_t and V_f, in the inner of the two groups and with respect to old normal subjects, O_i. Usually, the reconstructed dimension may be indicative of the number of basic variables that are involved in the system under consideration. The obtained results indicate that young subjects show differences with respect to old subjects relative to the number of basic variables involved but such differences are rather moderate. In the case of the two investigated pathologies we are in presence of a very different dynamics and attractor features in the inner of the groups relative to controls. All the arising differences lead to an interpretation in terms of a profound modification and alteration and of a more marked complexity of the dynamics in the V_t and the V_f cases compared to normal subjects. The results indicate that in some cases a larger number of variables while in other cases a smaller number of variables is required. This is indicative of the profound alteration that the two pathologies induce in heart dynamics, compared to the cases of normal subjects.

The second step was to calculate the largest Lyapunov exponent. For brevity, we avoided calculating the whole Lyapunov spectrum. The results are reported in Table 2. All the subjects gave positive values for the exponent. This may be indicative of the presence of chaotic regimes. It may be seen that young subjects gave values trending higher compared to old subjects. Signals immediately before Ventricular Fibrillation gave discordant results in the sense that in one case we had the lowest value of the Lyapunov exponent of the whole experimentation but we had also cases with values very similar to the high values that were obtained in the case of young subjects. On the contrary, signals immediately before Ventricular Tachycardia gave rather low results in two cases. The other remaining values are similar to those previously obtained for the old healthy subjects. In conclusion we had also in this case (as well as in the case of phase space reconstruction) a net variability in the results in the case of pathologies and a rather constant behaviour of λ_E in the case of normal subjects. In our interpretation these results confirmed that the investigated pathologies induce a profound modification and alteration in the dynamics of the two investigated processes compared to normal cases. The statistical results are given also in Table 2.

It is seen that we have significant differences in the case of young subjects vs. old subjects. These are interesting results since, as also outlined in previous papers by other authors [21], this means that the new paradigmatic rule in dynamics of R-R signals is its variability. Young subjects demonstrate a dynamics of R-R intervals that is based on a greater variability compared to old subjects. Age effects on beat-to beat fluctuations in human interbeat intervals involve a progressive reduction of variability and, in accord, we find a statistically significant difference in Largest Lyapunov Exponent between young and old subjects. We find also statistically significant differences in old subjects about thirteen minutes before the advent of an episode of ventricular tachycardia. This is a remarkable result. In fact, it says that we have an index, λ_E, that

Journal of Consciousness Exploration & Research| December 2010 | Vol. 1 | Issue 9 | pp. 1-69
Conte, E., Todarello, O., Conte, S., Mendolicchio, L., Mendolicchio, L. & Federici, A.
Methods and Applications of Non-Linear Analysis in Neurology and Psycho-physiology

17

is able to inform us in advance on the future advent of a so severe an episode in human heart dynamics. Unfortunately, such predictive value is not obtained also in the case of the Ventricular Fibrillation which in fact does not show significant differences compared to the case of old subjects. Other details are given in Table 2.

As third step of our analysis, we must now examine the structure of the investigated signals, and this kind of analysis may be performed by employing the RQA. Let us remember that we calculate a Recurrence Plot and the following variables of interest: the %Rec, the %DET, the %Lam, the T.T., the Entropy, the MaxLine, the Trend. Modifying slightly our previous language, we may reconsider here some of the variables. In particular, the recurrence rate estimates the probability of recurrence of a certain state. Stochastic behaviours cause very short diagonals while deterministic behaviours determine longer diagonals. Consequently, the ratio of recurrence points forming diagonals to all recurrence points, estimates the determinism. Diagonal structures show the range in which a part of the trajectory is rather close to another one at a different time. Therefore, the diagonal length is the time span they will be close and their mean represents the mean prediction time. The inverse of the maximal length line may be interpreted as the maximal positive Lyapunov exponent. The entropy is defined as the Shannon entropy in the histogram of diagonal line lengths. We may also compute the ratio between the recurrence points forming a vertical structures and the whole set of recurrence points. This variable is called Laminarity, and it related to the amount of laminar states and intermittency. In dynamical systems, intermittency is the alternation of phases of apparently periodic and chaotic dynamics. This is useful for the study of transitions (chaos-ordered, or chaos-chaos transitions). (TT), which is the mean length of vertical lines, measures the mean time that the system is trapped in one state or change only very slowly.

This is the basic scheme of RQA. We see that by this set of variables, we may actually explore the inner structure of the given signal, and this is the reason because RQA is so important in analysis of non linear dynamics in signals.

We may now return to consider the specific cases under our investigation. We performed the RQA analysis using a Radius R=20 so to maintain %Rec about 2-4% . This is a methodological attitude that is often usefull in such analysis. We selected a Line L=3, and we used Euclidean distance and mean rescaling. In Figures 5, 6, 7, 8 we give an example of recurrence plot for a subject of Y_i, O_i, V_{ti}, V_{fi}, respectively. The results of the RQA investigation are given in Table 3.

In Table 4 we have instead the statistical analysis of the RQA results.

Before inspection of recurrence plots, we remember the meaning of diagonal lines and in particular the fact that square areas, really a combination of vertical and diagonal lines, indicate laminar areas, intermittency, possibly suggesting transitional regimes as previously discussed. Still, let us observe that in Tables 3 and 4 we introduced a new variable, the Ratio = %Det / %Rec. We see that the signals employed in the investigation have actually a different inner structure. As expected, young subjects give statistically significant different results compared to old subjects for Laminarity, Trapping Time, Entropy, and Max Line. In brief, the two kinds of signals have a very different structure. Statistically significant differences are obtained also in the case of R-R of old subjects compared to R-R of old subjects before the advent of ventricular

Journal of Consciousness Exploration & Research| December 2010 | Vol. 1 | Issue 9 | pp. 1-69
Conte, E., Todarello, O., Conte, S., Mendolicchio, L., Mendolicchio, L. & Federici, A.
Methods and Applications of Non-Linear Analysis in Neurology and Psycho-physiology

18

tachycardia, and this happens for Determinism, Laminarity, Entropy, and MaxLine. This is a remarkable result since by it we are in the condition to anticipate the event. We may predict in advance the advent of ventricular tachycardia. Still, statistically significant differences are obtained in the case of old subjects compared to old subjects with ventricular fibrillation. In this case it is the Ratio variable this is indicative.

Finally, we observe that the structure of the two signals, one before the advent of ventricular tachycardia and the other before the advent of ventricular fibrillation, show significant differences, and this happens for Determinism, and Laminarity. Also this last result is remarkable since it suggests that we have two profoundly different pathologies that may be better studied and understood on the basis of such two variables.

This last observation completes our RQA investigation. In conclusion, we have given a number of important results relating the different structure and the dynamics of the signals under investigation. They all show relevant features that certainly will not fail to be studied and interpreted with care in their proper physiological and clinical context.

Let us conclude with the results obtained by our CZF method. Following the CZF methodology, we calculated the variogram using 1021 lags. On this basis we evaluated the most important parameter of the method, that is, the Total Variability (VT) of the given time series. It was expressed as the square root ot the total variability of the signal obtained for each lag. Therefore the results are expressed in sec. We also calculated the variogram distribution in the frequency domain, in substitution of the classical Fourier transform. Thus, we calculated the variability of R-R in sec^2 in the three bands of interest, VLF, LF, and HF. The results are given in Table 5. In Table 6 we give the results for statistical analysis (t-Test) and in Table 7 those for correlation analysis.

First, let us comment the Total Variability, VT. In the cases under investigation, it shows that young subjects, as expected, show a greater VT compared to old subjects. This parameter increases remarkably in R-R time series before the advent of ventricular fibrillation and of ventricular tachycardia. The statistical analysis reveals that we have a very significant difference in young subjects with respect to old subjects, and, particularly, in old subjects with respect to those with future ventricular fibrillation and in those with future ventricular tachycardia. We may conclude to have found an excellent predictive parameter that is able to anticipate the advent of severe events in hearth dynamics. In addition, we find also that statistically significant differences are maintained for VT in young subjects compared to old subjects for VLF, LF and HF bands in the frequency domain. Still, significant differences are found in old subjects compared to subjects with future ventricular tachycardia for LF and HF bands.

In order to go on in the understanding of such complex phenomena relating pathologies, we have also performed a correlation analysis finding other remarkable results. In young subjects VT results correlated in a significant manner with VLF, LF, and HF. In itself, this result does not appear to be so relevant. It becomes of particular interest when we consider also the results of correlation analysis for old subjects. In fact, in this case we obtain that the total variability of the signals correlates with LF and HF but not with with VLF. This is a very interesting conclusion that deserves to be explained and interpreted in detail under the physiological and clinical

Journal of Consciousness Exploration & Research| December 2010 | Vol. 1 | Issue 9 | pp. 1-69
Conte, E., Todarello, O., Conte, S., Mendolicchio, L., Mendolicchio, L. & Federici, A.
Methods and Applications of Non-Linear Analysis in Neurology and Psycho-physiology

19

profiles. Finally, we obtain still results of particular significance when we apply the correlation analysis to the case of future ventricular fibrillation and of future ventricular tachycardia. In fact, in the case of future ventricular fibrillation we find that correlation maintains between VT and VLF, between VT and LF, and between VT and HF, but in the case of ventricular tachycardia, correlation maintains only between VT and VLF, and between VT and VLF/(LF+HF). These results evidence in a quantitative manner the profound alterations that intervene in health dynamics soon before the advent of ventricular fibrillation and of ventricular tachycardia but also clear in detail the substantial differences that characterize the two pathologies. Certainly, there is here matter for physiologists and clinicians to find a proper understanding and interpretation of such results giving new insights in this matter.

To complete the present section we must still add something about the fractal dynamics of the investigated R-R time series. We previously outlined that a Generalized Fractal Dimension may be calculated by employing the CZF method. Otherwise, and for reason of brevity, we calculated in this section the Hurst exponent. The results are given in Table 8 where is also presented the statistical analysis. Also the analysis of Hurst exponet furnishes relevant results. Using this methodology, it is found that ventricular fibrillation and ventricular tachycardia profoundly modify health dynamics just before of their advent. In fact, by inspection of Tables 8, we see that the values of the Hurst exponent all remain under the value of 0.5, and this result shows that the regime of such R-R time series is of antipersistence and thus of absence of long range correlation. In addition we see that we have statistically significant differences between values in young and old subjects. In addition, very significant differences are found between old subjects and old subjects with future ventricular tachycardia. At the same time very significant differences hold also between old subjects and those with future ventriculat fibrillation. Therefore, we obtain an excellent parameter of prediction of future severe failure in heart dynamics. The reason for such results is that the advent of the mentioned pathologies profoundly alters the fractal structure of the signals taken in consideration in V_t and in V_f. In conclusion, our analysis offers an excellent set of parameters that may be considered as predictive of ventricular tachycardia and ventricular fibrillation.

4.5 The Application of the CZF Method in Analysis of Spontaneous EEG

We called this method as CZKF because its formulation was enriched also by the contributions of another author [17]. We are accustomed to analyse brain patterns of subjects by standard methodologies. Specifically, subjects are instructed to close their eyes and relax. Brain patterns are recorded as wave shapes that commonly show sinusoidal like behaviour. They are measured from peak to peak with a normal ranging from 0.5 to 100 μV. EEG records may be obtained by positioning 21 or more electrodes on the intact scalp and thus recording the changes of the electrical field within the brain. Generally, even up to 128 and more EEG channels can be displayed simultaneously and each corresponding to a standard electrode position on the scalp. The results of EEG signals are usually registered as voltage differences between pairs of electrodes with bipolar leads or between an active electrode and a suitably constructed reference electrode.

The problem in analysing EEG is to provide a proper method to extract its basic quantitative features by accurate procedures. The research regarding the methodology began more than 70

Journal of Consciousness Exploration & Research| December 2010 | Vol. 1 | Issue 9 | pp. 1-69   20
Conte, E., Todarello, O., Conte, S., Mendolicchio, L., Mendolicchio, L. & Federici, A.
Methods and Applications of Non-Linear Analysis in Neurology and Psycho-physiology

years ago. The basic tool was, and still remains Fourier analysis. The brain states of subjects demonstrate some dominant frequencies; namely:

1) *beta waves (12-30 Hz)*
2) *alpha waves (8-12 Hz)*
3) *theta waves (4-8 Hz)*
4) *delta waves (0.5-4 Hz)*

Over the last two decades the traditional Fourier analysis has been enriched by other methods, including the widespread application of time-frequency methods for signal analysis such as the Wavelet Transform (WT), and the Hilbert transform. These applications have enjoyed varying results. Because of its simplicity, Fourier analysis has dominated and still dominates data analysis efforts. Despite this, as it was outlined in the previous sections, it should be widely recognized that the Fourier transform assumes crucial restrictions which are often violated also in the EEG time series.

The consequences of improper FFT use are significant: the resulting spectrum will make little physical and physiological sense. The brain has an average density of about 10^4 neurons per cubic mm. Neurons are mutually connected into neural nets through synapses. Subjects have about 500 trillion (5×10^{14}) synapses, and the number of synapses per one neuron increases with age while the number of neurons decreases with age. Thus although rather structurally simple, the interconnections produce one of the most massive (functional) structures existing in nature. The natural way to think of this structure is that of a dynamic system governed by laws of non linearity and of non stationarity. We are in presence of a very complex system that again shows a great variedness and variability. Consequently, any method of analysis must quantify these features in order to generate valuable results. To this purpose we propose the CZKF method.

Obviously, the basic feature of the CZKF method is that by it we must estimate the variability that one has in the EEG for each band in a given time interval. This represents the new and important feature of the method. By CZKF we have the opportunity for the first time to evaluate with accuracy the variability of EEG in each of the bands characterizing the brain waves of interest. In this case we will express total variability in microvolts, obviously.

Consider an EEG sampled at 250 Hz. First of all we will calculate the variogram for different lags, *h*, as previously explained in detail. We will realize a diagram in which we have the values of the variogram in y-axis (ordinate) and correspondingly the $h - lag - values$ on the x-axis (abscissa). Soon after the step will be that one of a conversion of variogram values from time to frequency domain.

We proceed in the following manner:

$$\frac{1}{250 Hz} = 0.004 \sec$$

In this manner

Journal of Consciousness Exploration & Research| December 2010 | Vol. 1 | Issue 9 | pp. 1-69   21
Conte, E., Todarello, O., Conte, S., Mendolicchio, L., Mendolicchio, L. & Federici, A.
Methods and Applications of Non-Linear Analysis in Neurology and Psycho-physiology

$$\frac{1}{0.004(lag - h - value = 1)}$$

will represent the frequency with the corresponding value of variability at 250 *Hz*.

Similarly,

$$\frac{1}{0.004 \times 2(lag - value)} = 125 Hz$$

will represent the value of variability at 125 *Hz*, and so on for lag values $h = 3,4,5,\ldots$. In this manner we may reconstruct the variability of the EEG time series data as a function of the frequency.

Analysis of brain waves will be performed by integration of the calculated variability in each of the four groups of brain waves previously reported summing for each characteristic frequency band. In this manner we will estimate also

$$P(f) = 1/f^{\beta}$$

This last discussion completes the exposition of some features of our method. It may be applied to EEG as well as to ERP. In our previous papers [17], we examined eight normal subjects (5 female and 3 male with age ranging from 21 to 28 years old). All the subjects were at rest, watchful but with closed eyes. The sampling frequency was at 250 Hz.

We focused our analysis on the following electrodes: CZ, FZ, O2, and T4. Phase space reconstruction is useless in our case since we had the electrodes positioned on the scalp and their space separation corresponds to time delay. We used the Euclidean Norm that is the time series reconstructed as

$$\sqrt{x_{C_Z}^2(t) + x_{F_Z}^2(t) + x_{O_2}^2(t) + x_{T_4}^2(t)} = X_{EEG}(t)$$

and we calculated the variogram of $X_{EEG}(t)$ at the various lags and subsequently the results were converted into *Hz*. 30000 points of EEG were used, corresponding to 2 minutes of recorded brain activity.

The results are reported in Fig. 9 and in Table 9. It gives an accurate reconstruction of the variability of brain activity in the four bands of interest that are the beta, alpha, theta and delta brain waves. Obviously the method fully substitutes the less appropriate application of FFT, Wavelet, Hilbert transformations and other linear applications.

Finally, we aim to outline here the interest of our CZKF method also in applications in cognitive studies, in analysis of IQ or also, for example, and to evaluate the anesthetic adequacy. In this manner our approach links the previous fundamental studies that are currently conducted by El

Journal of Consciousness Exploration & Research| December 2010 | Vol. 1 | Issue 9 | pp. 1-69   22
Conte, E., Todarello, O., Conte, S., Mendolicchio, L., Mendolicchio, L. & Federici, A.
Methods and Applications of Non-Linear Analysis in Neurology and Psycho-physiology

Naschie [22] and by Weiss H. and Weiss V [23]. As the CZKF method evidences in detail, the variance of the EEG may be quantified, and is a function of its frequencies. It becomes possible to scale and to measure inter-individual differences – for level of cognition, IQ or anesthetic adequacy not by any absolute score, but by the inter-individual variance of the subjects. Weiss and Weiss [23], in particular, based on empirical data of different authors, showed that thinking can be understood, if we see thoughts as macroscopic ordered (quantum) states in the sense of statistical mechanics. Thinking seems only to be possible, if brain waves use the mathematical properties of the golden ratio and hence of fractal-Cantorian spacetime as discussed by El Naschie [22]. Therefore, a straightforward application of the method and measure here developed is to test the IQ of subjects and correlate the measures arising from CZKF with IQ, using power and variance in the entire range from 3 to about 30 Hz of the EEG.

5. An Analysis of State Anxiety

5.1 Introduction

We will develop now a final application. We will study the state anxiety in humans. We will apply all the previous exposed methodologies. In order to delineate in detail such developed research, we retain that we will help the reading exposing this argument avoiding any possible intermixture with the previous ones, and thus separating this argument from the previous ones, using also references, tables and figures that relate a separate and independent numeration respect to the previous one, used to illustrate the general field of methodologies and applications.

Let us start with a brief discussion on the use of non linear methodologies in psychology.

Psychological data were usually collected in the past psychological studies to assess differences between individuals or groups which were considered to be stable over time (1,2). Instead, a further approach has gained relevance in the past decade, which is aimed to perform an intensive time sampling of psychological variables of individuals or groups at regular intervals, to study time oscillations of the collected data (1). In this way human behavior has been investigated to analyze, for instance, the impact of everyday experience on well-being (3) or the after-effects of negative events (4) or to examine the association between emotions and behavioral settings (5). These studies were often aimed to analyze the nature of rhythmical oscillations in mood and performance of human beings (6). Such an approach leads to progressive changes not only in the methods to sample psychological data but also in our way of thinking about many psychological variables, which may be considered as expression of mind entities unfolding over time (1). A reason to outline the importance of this approach is to acknowledge the role of the human interactions in governing the transitions which continuously take place in mind entities.

It is becoming relevant the notion that our mind, our ideas and convictions are all formed as the results of interactive changes and all they follow possibly a quantum like behavior. Let us explain in detail what we mean by this statement (7,8). For certain questions, individuals have predefined opinions, thoughts, feelings or, still, behaviors. This kind of condition may be considered to be stable in time in the sense that an intensive time sampling of data, consisting as example to questions asked to an individual from an outsider observer or by himself at regular

Journal of Consciousness Exploration & Research| December 2010 | Vol. 1 | Issue 9 | pp. 1-69   23
Conte, E., Todarello, O., Conte, S., Mendolicchio, L., Mendolicchio, L. & Federici, A.
Methods and Applications of Non-Linear Analysis in Neurology and Psycho-physiology

time intervals, will simply record a predefined answer that never will be determined and actualized at the same time the question is posed .In this case, we have a stable dynamic pattern for individuals or groups. The intensive time sampling of data will only confirm an information on time dynamics that is stable in reporting a pattern in self-report or in performance measures with regard to behavior in time of the involved individuals. It has been evidenced (7, 8) that, under the profile of a statistical analysis, the cases as those just mentioned, in which individuals have a predefined opinion or thought that may not be changed in time at the same moment in which questions are actually posed, correspond to a kind of classical dynamics that, statistically speaking, may be analyzed in terms of classical statistical approaches since they are not context dependent (7,8). There are situations in which, instead, a person, who is being questioned by himself or by an outsider observer, has no predefined opinion or thought or feeling or behavior on the given question. The kind of opinion, as example, is formed (that is to say: it is actualized) only at the moment in which the question itself is posed and it is formed on the basis of the context in which the same question is posed. This is a case of a quantum like behavior for a cognitive entity. The core of the difference resides in the fact that in the case of quantum like behavior we are dealing with the actualization of a certain property that is dependent from the instant of time in which the question is posed and thus, in particular, it depends also from the context in which it is posed while, instead, in the classical case all properties are assumed to have a definite connotation before the question itself is posed and thus they are time and context independent. Processes of the first kind are said quantum like, and they follow a quantum like statistics (7, 8). The basic content of such quantum probability approach is the calculation of a probability of actualization of one among different potentialities as result of the individual inspection itself or of an outsider observation. New paradigms are thus emerging in studies regarding mind behavior: one is the concept of potentiality, linked to the concept of actualization. Still, we have the concept of dynamic pattern that is linked to the observation of changing in time as result of the interactive transitions (potentiality-actualization) which take place in human interactions. The case of quantum like behavior is one of the manifold situations in which an intensive time sampling of psychological data, may give important information on the dynamic patterns in self-report and performance measures.

It is noteworthy that people have a defined "sense of self" and accompanying memories of a very early age. It may be due to the fact that the "attractor" of personality (as developed by the brain) has not established a defined enough probability of neuronal connections to establish such a distribution: if neuronal connections are essentially uniform in their shape, it is questionable if an attractor is defined. With repetitive learning inputs, the probability distributions become established (narrowed) and "personality" emerges. Learning skills proceeds along similar lines: repetitive "habits" further narrow the probability distributions so as to make a particular action more refined to the point of not requiring active effort. Both personality and learning, however, are dependent upon the genetics which establish the basic physiology of the neuronal machinery. Predictability regarding personalities and activity is by definition of the singular dynamics, a stochastic process: no matter how narrowed the probability distributions, there always remains a level of uncertainty.

The performance of current neural networks is still too "rigid" in comparison with even simplest biological systems. This rigidity follows from the fact that the behavior of a dynamical system is fully prescribed by initial conditions. The system never "forgets" these conditions: it carries their

Journal of Consciousness Exploration & Research| December 2010 | Vol. 1 | Issue 9 | pp. 1-69   24
Conte, E., Todarello, O., Conte, S., Mendolicchio, L., Mendolicchio, L. & Federici, A.
Methods and Applications of Non-Linear Analysis in Neurology and Psycho-physiology

"burden" all the time. In contrast to this, biological systems are much more flexible: they can forget (if necessary) the past, adapting their behavior to environmental changes.

The thrust here is to discuss the substantially new type of dynamical system for modeling biological behavior introduced as non deterministic dynamics. The approach is motivated by an attempt to remove one of the most fundamental limitations of current models of artificial neural networks—their "rigid" behavior compared to biological systems. As has been previously exposed in detail, the mathematical roots of the rigid behavior of dynamical systems are in the uniqueness of their solutions subject to prescribed initial conditions. Such an uniqueness was very important for modeling energy transformations in mechanical, physical, and chemical systems which have inspired progress in the theory of differential equations. This is why the first concern in the theory of differential equations as well as in dynamical system theory was for the existence of a unique solution provided by so-called Lipschitz conditions. On the contrary, for information processing in brain-style fashion, the uniqueness of solutions for underlying dynamical models becomes a heavy burden which locks up their performance into a single-choice behavior.

A new architecture for neural networks (which model the brain and its processes) is suggested which exploits a novel paradigm in nonlinear dynamics based upon the concept of non-Lipschitz singularities [7, 8]. Due to violations of the Lipschitz conditions at certain critical points, the neural network forgets its past as soon as it approaches these points; the solution at these points branches, and the behavior of the dynamical system becomes unpredictable. Since any vanishingly small input applied at critical points causes a finite response, such an unpredictable system can be controlled by a neurodynamical device which operates by noise and uniquely defines the system behavior by specifying the direction of the motions in the critical points. The super-sensitivity of critical points to external inputs appears to be an important tool for creating chains of coupled subsystems of different scales whose range is theoretically unlimited.

Due to existence of the critical points, the neural network becomes a weakly coupled dynamical system: its neurons (or groups of neurons) are uncoupled (and therefore, can perform parallel tasks) within the periods between the critical points, while the coordination between the independent units (i.e., the collective part of the performance) is carried out at the critical points where the neural network is fully coupled. As a part of the architecture, weakly coupled neural networks acquire the ability to be activated not only by external inputs, but also by internal periodic rhythms. (Such a spontaneous performance resembles brain activity). It must be stressed, however, that behavior may be predicted in the sense of establishing a probability distribution of choices. Thus behavior is not determined, but 'guessed' within the bounds of the probability distribution.

In its most simple form, consider, for example, an equation without uniqueness:
$$dx/dt = x^{1/3} \cos \omega t.$$
At the singular solution, $x = 0$ (which is unstable, for instance at $t = 0$), a small noise drives the motion to the regular solutions, $x = \pm (2/3\omega \sin \omega t)^{3/2}$ with equal probabilities. Indeed, any prescribed distribution can be implemented by using non-Lipschitz dynamics. It is important to emphasize, however, the fundamental difference between the probabilistic properties of these non-Lipschitz dynamics and those of traditional stochastic or differential equations: the

Journal of Consciousness Exploration & Research| December 2010 | Vol. 1 | Issue 9 | pp. 1-69   25
Conte, E., Todarello, O., Conte, S., Mendolicchio, L., Mendolicchio, L. & Federici, A.
Methods and Applications of Non-Linear Analysis in Neurology and Psycho-physiology

randomness of stochastic differential equations is caused by random initial conditions, random force or random coefficients; in chaotic equations small (but finite) random changes of initial conditions are amplified by a mechanism of instability. But in both cases the differential operator itself remains deterministic. Thus, there develops a set of "alternating," "deterministic" trajectories.

We would now discuss the reason of a terminology that is delineating. As said, the analogy is with the physics. The state s(t) of a physical entity S at time t represents the reality of this physical entity at that time. In the case of classical physics the state is represented by a point in phase space while in quantum physics it is represented by a unit vector in Hilbert space. In classical terms the state s(t) of the physical entity S determines the values of all the observable quantities connected to S at time t. The state q(t) of a quantum entity is represented instead by a unit vector of Hilbert space, the so called normalized wave function $\psi(r,t)$. For a quantum entity in state $\psi(r,t)$ the values of the observable quantities are potential: this is to say that a quantum entity never has, as example, simultaneously a definite position and a definite momentum and this represents the intrinsic quantum indeterminism that affects reality at this level. We have the relevant concept of potentiality: a quantum entity has the potentiality to realize some definite value for some of its observable quantities. This happens only at the moment of the observation or of measurement and it is this mechanism that realizes a transition from a pure condition of potentiality to a pure condition of actualization. A definite value is not actually realized in the potential state $\psi(r,t)$. A definite values is really actualized only at the moment of the direct observation of some property of the given entity and through the same mechanism of the observation during the act of the measurement. The novel feature is in the transition potentiality → actualization that characterizes the mechanism of observation and measurement.

We have to realize here a large digression in order to clear in detail this point that appears to us of fundamental importance.

As we know all quantum mechanics is based on such binomial conceptualization of potentiality from one hand and actualization from the other hand. In particular, the actualization corresponds to the observation and measurement or, that is to say, to the moment in which we become conscious that some kind of measurement has happened (collapse of wave function) since we read its result by some device. Generally speaking, a system is in a superposition of possible states (superposition principle, potentiality) and such superposition principle is violated in a measurement. This led von Neumann to postulate that we have two fundamentally different types of time evolution for a quantum system. First, there is the casual Schrödinger equation evolution. Second, there is the noncasual change due to a measurement and this second type of evolution (passage from potentiality to actualization) seems incompatible with the Schrödinger form. This situation forced von Neumann to introduce what is usually called the von Neumann postulate of quantum measurement. This happened about 1932. Rather recently, one of us (EC), using two theorems in Clifford algebra, has been able to give a complete justification of von Neumann postulate. The result has appeared on International Journal of Theoretical Physics, and it is available on line [8]. Thus we have given proof of a thing that for eighty years remained a postulate, often discussed and largely questioned. This new result, at least under an algebraic profile, explains the wave function collapse and gives total justification of it, also giving to quantum mechanics an arrangement as self-consistent theory that in the past was often

Journal of Consciousness Exploration & Research| December 2010 | Vol. 1 | Issue 9 | pp. 1-69
Conte, E., Todarello, O., Conte, S., Mendolicchio, L., Mendolicchio, L. & Federici, A.
Methods and Applications of Non-Linear Analysis in Neurology and Psycho-physiology

26

questioned as missing in the theory and signing such missing as a probe of weakness of such theory. In conclusion, the passage potentiality – actualization now seems a more demonstrated transition to which we have to attribute the greatest importance if we do not aim to remain linked to a too limited vision of our reality. On the other hand, there is no matter to continue an infinite discussion on a possible link between quantum mechanics and cognition. We have unequivocal results that demonstrate in detail such point. It is universally accepted that J. von Neumann showed that projection operators represent logical statements. In brief, J. von Neumann showed that we may construct logic starting from quantum mechanics. According to the fundamental papers published by the great logician Yuri Orlov, and in the light of the results that, we repeat, one of us has recently obtained, it may be unequivocally shown that also the inverted passage is possible. Not only we may derive logic on the basis of quantum mechanics. We may derive quantum mechanics from logic. So, the ring is closed. The link between quantum mechanics and cognition is strongly established. The split that occurred between psychology and the physical sciences after the establishment of psychology as an independent discipline cannot continue to encourage a delay in acknowledging this thesis. We may be convinced that there are levels of our reality in which the fundamental features of logic and thus of cognition acquire the same importance as the features of what is being described. Here we no more can separate "matter per se", in Orlov words, from the features of logic and cognition used to describe it. We lose the possibility of unconditionally defining the truth, as we explained previously, since the definition of truth, now depend on how we observe (and thus we have cognition) the physical reality . Obviously such relativism does not exist in classical mechanics while instead by quantum mechanics we have a Giano picture able to look simultaneously on the left and on the right, at cognitive as well as physical level.

Let us return now to the central problem we have in discussion.

Some mind entities follow quantum like behavior (7, 8). Let us restrict our example to the case of a cognitive entity. A psychological task asks to a participant a question that has a predefined value as answer for each individual. The task asks, as example, to the participant if he (she) has blue eyes. It is clear that the cognitive entity of the participant has a predefined opinion on this question and the measurement, corresponding to the act of posing the question to the subject, will furnish only the trivial recording of an output that is predefined also before the question is posed. There are cases in which the cognitive entity may be submitted to a question for which the person who is being questioned has no opinion ready. He has several potentialities and only one of such potentialities will be actualized at the moment the question is being asked. As example, let us admit that the posed question is the following: are these two geometrical figures equal? (an ambiguous figure). At the moment the question is being asked, the subject has no predefined opinion. He may have, as example, two potential states (possibilities) that are superimposed and they are the two possible answers: yes and no. The cognitive entity will actualize only one answer among the two possible ones at the moment the question is posed and such actualization will correspond to an act of consciousness of the subject. Through the posed question, the subject will be induced to a transition from to a condition of potentiality to that one of actualization. In a quantum like framework, such mechanism of transition from potentiality to actualization will be intrinsically stochastic and strongly dependent from the context in which the cognitive entity of the subject will be induced to answer. Potentiality states of mind entities are superposition of potentialities that are characterized at an ontological level and, as said, among the different

Journal of Consciousness Exploration & Research| December 2010 | Vol. 1 | Issue 9 | pp. 1-69   27
Conte, E., Todarello, O., Conte, S., Mendolicchio, L., Mendolicchio, L. & Federici, A.
Methods and Applications of Non-Linear Analysis in Neurology and Psycho-physiology

potentialities only one state will be actualized corresponding to an act of introspective activity (consciousness advent) of a subject. It is clear that in such cases an intensive time sampling procedure enables to collect data relating subsequent individual acts of introspection, of actualization, of conscious aware and this represents an interesting technique for analysis of mind dynamics.

It is important to outline here that the approach of using an intensive time sampling of psychological data is relevant not only in the cases in which a quantum like behavior may be assumed but, generally speaking, in all the cases of experimentation in which there is the reasonable motivation to retain that it is the dynamic evolution in time of mind entity to cover an important role in the framework of the investigated phenomenology.

It remains to evidence that, through an intensive time sampling of psychological data, we realize a discrete collection of results that usually we call a time series of data. They are actually used extensively in physiological studies of biological signals, and the importance is related to the fact that they contain a fingerprinting of the process under investigation. Consequently, the basic finality of this kind of studies is to analyze the nature of the observed fluctuations in time. Generally, the analysis of the data may enable to establish relevant questions as if time evolution follows a linear or a non linear dynamics, and in particular if it is regulated by deterministic, or chaotic deterministic or noise influenced patterns.

In the present study we investigated the phenomenon of anxiety of state. The finality was to introduce new parameters for the interpretation and control of such psychological manifestation.

5.2 The Phenomenon of Anxiety

Anxiety may represent a proper condition to investigate in detail potentiality of mind entities in analysis of time dynamic pattern. It is well known that fear is profoundly distinguished from anxiety. It is known from many models (9), that fear is a response to a present and actual danger while, generally speaking, anxiety is a response to a potential danger. According to our quantum like model of the previous section, we may say that the anxious individual, at fixed times, may give his conscious introspection and thus evaluating and actualizing a danger that is only potentially fixed. Therefore, fear is a response to a present - real danger, anxiety is instead the response to a potential danger. In various models (10) the risk assessment is seen as the central component of anxiety and it is realized in terms of approaching and scanning potentially dangerous situations. Fear and anxiety can each produce a physiological arousal response that involves activation of the adrenergic system in the CNS and in sympathetic branch of the autonomic nervous system (SNS) (11). Since such identical systems are involved in both such conditions, the phenomenological experience of arousal seems similar, and such similarity of arousal experiences contributes to the common tendency to retain fear and anxiety as either interchangeable manifestations. There are instead substantial differences. The central difference between fear and anxiety should reside in the kind of quantum like behavior that we established in the previous section. The individual in a state of fear perceives the threat that is immediate and real and, on this basis, he gives an active response that in some manner is just induced from the external stimulus. In other terms, the individual actualizes a response that, in some sense, is defined on the basis of the kind of real perceptive stimulus that is offered to him. In the case of anxiety, the individual does not perceive an immediate threat (there is not an external stimulus

Journal of Consciousness Exploration & Research| December 2010 | Vol. 1 | Issue 9 | pp. 1-69   28
Conte, E., Todarello, O., Conte, S., Mendolicchio, L., Mendolicchio, L. & Federici, A.
Methods and Applications of Non-Linear Analysis in Neurology and Psycho-physiology

that actualizes the response). He is focused on a potential threat for the immediate or future times and in many cases he inherits this condition on the basis of his personal history and psychological background (see, as example, the case of a subject with post traumatic stress disorder). In analogy with intrinsic quantum indetermination of physical reality, there is here a proper condition of quantum like uncertainty for mind entity: owing to the indeterministic nature of the anxiety-producing threat, the individual remains suspended into potential states, and usually he cannot determine whether to act or how to act. This is the clear indication of quantum like behavior. The anxiety-producing threat is only potential: the individual feels that there is something that may happen or that might not happen; he remains suspended in a superposition of such potential states. He continues to think about the threat (he remains in the superposition of potential states). He does not react to an attack or to a perception of being attacked, but he remains in the suspended possibility of being attacked. If such individual perceives himself to be really under an attack (actualization), then he will enter an actual fear state.

One very interesting feature is that anxiety represents an emotional condition that is so general and so radical in human that it cannot be considered only a sign of pathology or a defined syndrome but a general mode of the human existence with extreme values that obviously enter in the domain of psychopathology. Therefore, the time analysis of its dynamics offers an excellent opportunity to analyze basic features of a time dynamics regarding in general mind entities of human existence. In addition, while the anxiety of trait may be considered as a rather stable condition of our personality, the anxiety of state is considered more linked to transient phases of our everyday emotional condition, and it may be evaluated by using proper test that were introduced by C. D. Spielberger starting with 1964 (12). It is important to outline that the test may be repeated at fixed time intervals so to have a final time series of collected data that are indicative of the changing in time of the phenomenology under study. The individual is asked to answer to twenty fixed questions that were elaborated (12) with the direct finality to quantify the value of the anxiety of state at the moment of the administered test. For each question, the individual has at his disposal four different modalities of answer with a calibrated score ranging from 1 to 4 according to the seriousness of the emotional condition.

The value of state anxiety for each administered test to the individual, is usually evaluated by direct calculation of the achieved total score and, in case, a statistical analysis may be developed in order to obtain standard statistical indexes over a proper range of time. It is evident that this manner to proceed results to be very limited. We are certainly interested to the values of the test but mainly we must focus our attention on the manner in which variations and oscillations of test values are induced in time from mind entities. For this purpose, the introduction of new methodologies and parameters is really required in order to characterize the dynamic pattern of anxiety of state in individuals: such parameters should be useful also to elaborate diagnostic as well as therapeutic strategies. From the viewpoint of our quantum like model the results of the test must be conceived in the following manner: at fixed times, the individual, through each posed question, exerts an introspective activity on himself (an act of conscious awareness): by each introspective act the subject makes a transition from a superposition of four potential states (the four kinds of answer that are at his disposal) to the final actualization of only one among such four potentialities. None of the four potentialities is predefined previously the question is posed (superposition of potential states) and only one among the different potentialities is actualized only at the moment of the conscious introspection (transition potentiality-

Journal of Consciousness Exploration & Research| December 2010 | Vol. 1 | Issue 9 | pp. 1-69
Conte, E., Todarello, O., Conte, S., Mendolicchio, L., Mendolicchio, L. & Federici, A.
Methods and Applications of Non-Linear Analysis in Neurology and Psycho-physiology

29

actualization). Obviously, there exists also here a limit in our experimentation. When the subject repeat his(her) test for the second time, he just knows what question will be posed to him and this situation could influence his answer. However, we will admit here that the subject, a control subject not affected from pathologies, will be able to answer to the posed questions without suffering a strong conditioning arising from the fact that he previously knows the posed questions. Our aim is to investigate the nature of such transitions, potentiality-actualization, in time.

5.3 Materials and Methods

Six healthy subjects were examined: F. Dav., male, 30 years old, D.Pet., female, 25 years old, A.Mac., female, 55 years old, G.Den., female, 30 years old, A.Men., male, 57 years old, M. Den., female, 32 years old. Each subject was subjected to the test four times in one day and precisely at each time step of three hours starting with the waking up. The collection of data proceeded for about 30 days. Time series data were collected by the test given to each subject . The resulting time series data for the subject D.Pet. is reported in Fig.1 to give an example of the obtained experimental time dynamic pattern.

5.4   Results of Poincaré-plot Analysis of the Data

In this section, we aim to introduce new indexes that in our opinion may help in the characterization of the investigated process.

As usual, our examination of the data started with the elaboration of a statistical analysis for the six examined subjects. The results are reported in Tables 1-6 for each subject. Mainly, we calculated the mean, the standard deviation and the variance of the scores obtained in about 30 days. Of importance it must be considered the value of the variance since, as previously said, we were mainly interested to investigate the phenomenology of the variations, and thus of the oscillations and of the fluctuations of the test value during the time period of its administration. We added also some other statistical indexes as the Median, the Minimum-Maximum values, the Root Mean Squared, the Skewness and the Kurtosis to have a clear characterization of the correctness of our samples under a statistical profile. In fact, it may be verified by these indexes that all the subjects responded to the test with full adherence to the requirements of the correct statistical samples.

A subject reached a mean value of 23.4 for the test, the other reached 30.3, the subsequent obtained 38.1, the other had 38.3 and, finally, the two remaining subjects had 47.2 and 53.4 respectively. It is important to outline here that the test furnishes usually four different scales for the evaluation of the score, the first with score value of 20 (very moderate level of anxiety of state), the second with value ranging from 21 to 40 (moderate level of anxiety of state), the third with score value ranging from 41 to 60 (high level of anxiety of state), and the fourth with score value ranging from 61 to 80 (very high level of anxiety of state). Therefore, four subjects resulted to have a moderate level of anxiety of state, and the remaining two subjects resulted instead to have an high level of anxiety of state. Note that the use of the mean value of the test in time and the subdivision of the test score in four intervals does not help for a correct diagnostic identification of the dynamic pattern of each subject in time. Looking at the values of the

Journal of Consciousness Exploration & Research| December 2010 | Vol. 1 | Issue 9 | pp. 1-69   30
Conte, E., Todarello, O., Conte, S., Mendolicchio, L., Mendolicchio, L. & Federici, A.
Methods and Applications of Non-Linear Analysis in Neurology and Psycho-physiology

Standard Deviations and of the Variances for such subjects, one catches sight of profound differences among subjects included instead in the same interval. As example, in the case of two subjects we have mean values of 38.1 and 38.3 that are very similar under the profile of the mean score but they exhibit profound differences under the profile of their variability in time since one has a Standard Deviation of 6.39 and a Variance of 40.86 while the other subjects has a Standard Deviation of 9.99, and a Variance of 99.89. Therefore, it derives that in no manner the mean value of score in the test for anxiety of state may be assumed to represent the correct diagnostic profile of the anxiety of state of a subject in time. From the previous section we know that such dynamic profile could be quantum like and as such marked from pure stochastic behaviors. Therefore we must be interested to a very deep analysis of such time variations, as oscillations and fluctuations of state anxiety in time and to this purpose the introduction of proper new indexes is primarily required. Before of all, in order to proceed along an accurate characterization of the time dynamic of state anxiety of subjects, we aim to introduce two new indexes. They are obtained reconstructing a kind of phase space with x_i values in abscissa against x_{i+1} values in ordinate. On a fitted ellipse we identify two indexes, the first, that we call here SD1 in analogy with previous studies on heart rate variability, expresses the tendency to the variability of the score for each subject in the short time intervals, and the second, that we call SD2 for the same analogy, expresses instead the tendency to the variability in the score along a consistent time interval. The use of such both indexes, SD1 and SD2, gives us the manner to characterize and to examine the dynamic tendency of state anxiety for each subject along the time interval of the investigation, considering variations of this phenomenon in the brief interval of time as well as in the larger time interval.

Poincaré-plots (13) are currently employed to investigate the complex dynamics of non linear processes as those given in Fig.1. A two-dimensional phase space may be used to visualize the information contained in a given time series. In a Cartesian co-ordinate system a point P_i is defined by the time interval T_i and τ intervals subsequently following T_i, thus giving $P_i(T_i,T_{i+\tau})$, being τ a proper time delay that in studies of chaotic-deterministic time series may be estimated by using Autocorrelation Function and Mutual Information Function (14). In this our preliminary investigation a time delay $\tau=1$ was selected by us. In this manner, the Poincaré-plot resulted to be a diagram in which each data of the given time series is plotted as a function of the previous one ($\tau=1$), this plot gives a visual inspection of the given time series data by representing qualitatively with graphic means the kind of variations of such data fingerprinted during their collection. The realized plots may be analyzed also quantitatively. This quantitative method of analysis is based on the assumption of different temporal effects of changes on the subsequent time series data without a requirement for a stationary behavior of data itself. Analysis, generally, entails fitting an ellipse to the plot with its center coinciding with the center point of the markings. The line defined as axis 2 shows the slope of the longitudinal axis, whereas axis 1 defines the transverse slope that is perpendicular to axis 2. Usually, the Poincaré-plot is first round 45 degree ring, clockwise. The standard deviation of the plot data is then computed around the axis 2 and passing through the data center. The first index, SD1, is so calculated. SD1 accounts for the variability of the data for short intervals of time. The standard deviation of long term data is quantified by turning the plot 45 degree ring, counterclockwise, and by computing this time for data points around axis 1 which passes through the center of the data. SD2 is calculated and it accounts for variability of data for long term time intervals. In conclusion, given the time series data, we may introduce two indexes, SD1 and SD2 respectively,

Journal of Consciousness Exploration & Research| December 2010 | Vol. 1 | Issue 9 | pp. 1-69
Conte, E., Todarello, O., Conte, S., Mendolicchio, L., Mendolicchio, L. & Federici, A.
Methods and Applications of Non-Linear Analysis in Neurology and Psycho-physiology

31

that account for the variability of the analyzed data in short as well as long intervals of time. Applying this kind of analysis to our time series of data, we become able to estimate how is expected variability of state anxiety in subjects in short as well as in long intervals of times. The particular relevance of such two introduced indexes must be thus clear. It seems reasonable to conclude that subjects with low values of SD1 and SD2 will exhibit low levels of state anxiety while from a psychological and clinical viewpoint it will be carefully characterized the condition of subjects with high values of SD1 and SD2 or of SD1 and not SD2 or viceversa. A new phenomenology of state anxiety is so delineated, and it will be characterized by levels of state anxiety that will result to be discriminated and carefully characterized respect to the proper case of normality (low level of state anxiety). Statistically speaking, the plot will display the correlation possibly existing between consecutive data (scores of the test) in a graphical manner. Non linear dynamics considers the Poincaré-plot as a two dimensional reconstructed time series data phase space which is a projection of the reconstructed attractor describing, in our case, the dynamic of the mind entities responsible for state anxiety.

Concluding: The time series data of state anxiety will give a Poincaré-plot that typically will appear as an elongated cloud of points oriented along the line of identity. The dispersion of points perpendicular to the line of identity will reflect the level of short term variability of the score for state anxiety (SD1) while the dispersion of points along the line of identity will indicate the level of long term variability of the score for state anxiety (SD2). The elliptic structure will mirror instead the basic periodicity of the data and thus, as an important indication, it will correspond to the possible periodicity of the scores during the administered test.

We performed this analysis for the six subjects. The results of the Poincaré-plots are reported in Figures 2-7 while the quantitative results for SD1 and SD2, respectively, are given in Table 7 where they are compared with the mean values (s.d. and variances)of the scores of the test as they were previously calculated. The satisfactory predictive power of SD1 and SD2 is clearly evidenced.

Let us comment briefly some results. The subject F. Dav reported a mean value of 23.4 with st. dev. of 2.7 and a variance of 8.30. SD1 resulted 2.11 while instead SD2 reached the value of 3.74. This means that in the short time interval such subject varied his score of only 2.11(thus ranging in mean from 21.29 to 25.51) and thus remaining any way in a moderate level of state anxiety. In the long intervals of time his score changed of 3.74 and thus ranging in mean from 20.00 to 27.14 that is still low and very similar to variability in short time intervals. In conclusion this subject had a rather stable condition of moderate state anxiety.

The subject A. Men had a mean value of 30.3 with a st. dev. of 2.4 and a variance of 7.36. It resulted SD1=2.21 and SD2=3.13. The time dynamic of state anxiety of this subject seems to be very similar to that one of the previous subject with a rather stable tendency to remain in the condition of moderate level of anxiety in short as well in long intervals of times.

Let us consider now the case of A. Mac. who had a mean value of 38.1 with a st. dev. of 6.39 and a variance of 40.86. From the statistical data we deduce that his mean value is only of 7.8 points greater than A. Men. (corresponding to about 21%) but standard deviations and variances result to be very different. Owing to the great value of the variance we expect for such subject a great

Journal of Consciousness Exploration & Research| December 2010 | Vol. 1 | Issue 9 | pp. 1-69
Conte, E., Todarello, O., Conte, S., Mendolicchio, L., Mendolicchio, L. & Federici, A.
Methods and Applications of Non-Linear Analysis in Neurology and Psycho-physiology

32

tendency to variability and it is of importance to establish if such tendency to time variability regards the short or the long time intervals or both. The use of Poincaré-plot gives this kind of information. In fact, SD1 resulted to be 5.46 while instead SD2 gave the value of 7.33. Comparing such results with those of F. Dav and of A. Men., we conclude that the subject A. Mac. has a tendency to a great variability both in short as well in long time intervals. In short times his score may vary in mean ranging from 32.64 to 43.56 (he may reach also the high level of state anxiety), and in the long interval of time, his score may vary in mean from 30.80 to 45.40 (he may reach the high level of state anxiety also greater than ones of short time terms). In conclusion, this subjects has dynamic features that result to be very different from the previous ones. As the first two subjects, A. Mac. starts in mean with a moderate level of state anxiety but in the short time intervals as well as in the long time intervals he has the tendency to reach also high levels of state anxiety.

In conclusion, as seen, SD1 and SD2 compete in an evident manner to differentiate in detail the dynamic of state anxiety also for subjects that have very similar scores.

Let us examine now a very different situation. The subject D. Pet had a mean value of score of 38.3 with a st. dev. of 9.00 and a variance of 99.89. Note that really the mean value (38.3) of this subject is substantially the same (38.1) of the previous subject A. Mac. Profound differences exist instead for st. dev. and variances indicating that, in spite of very similar results for the test, the two subjects exhibited very different dynamic patterns that are important to characterize. In fact, calculating SD1 and SD2, we obtain that SD1=5.82 while SD2 actually assumes the value of 12.96. In comparison with A. Mac, the subject D. Pet. has a very similar value of SD1 (5.82 vs 5.46) but a very large difference for SD2 (12.96 vs 7.33). In the long time intervals this subjects presents a variability that may be also of about 13 times higher, confining him in a condition of high state anxiety. Therefore his dynamic pattern is very different from that one of A. Mac although the scores of the test resulted substantially the same (38.3 vs 38.1).

In addition the value of SD1 and SD2 for D. Pet. may be compared with the previous ones of F. Dav. that showed the most stable condition of moderate state anxiety . In this case, we may evidence a great suitability of SD1 (5.82 vs 2.11) and SD2 (12.96 vs 3.74) to actually characterize state anxiety of subjects and their variability in time. Looking at the results of Table 7 we may still comment the values of SD1 and SD2 that were obtained for the remaining subjects, G. Den and M. Den, observing that such indexes still continue to characterize in detail the time variability of state anxiety also for such subject.

In conclusion, we suggest that, in addition to the scores that are collected by the test, two other indexes should be adopted in order to proper characterize time variability and thus time dynamics of state anxiety of individuals and they are SD1 and SD2 as they are obtained by analysis of the obtained time series data by using Poincaré-plots.

5.5 Results of Variogram and Fractal Analysis

If SD1 and SD2 are two quantitative indexes that, as seen, are suitable to characterize the variability in short as well as in long time intervals for time series data regarding state anxiety, such indexes, of course, cannot give any detailed information on the time dynamics that

Journal of Consciousness Exploration & Research| December 2010 | Vol. 1 | Issue 9 | pp. 1-69   33
Conte, E., Todarello, O., Conte, S., Mendolicchio, L., Mendolicchio, L. & Federici, A.
Methods and Applications of Non-Linear Analysis in Neurology and Psycho-physiology

characterizes state anxiety. In order to reach this objective, a kind of non linear analysis must be still performed using some other elaborate techniques.

Let us start considering the notion of fractal. This term was introduced (15) in 1983 by B.B. Mandelbrot. A fractal object is made of parts that are similar to the whole in some way, either the same except for scale or statistically the same. The chaos dynamic mechanism and the interaction of non linear processes may be an essential cause of uneven distributions of data which results in fractal structure. Self-similarity or statistical self-similarity may be investigated in given time series data with the finality to establish their fractal or multifractal behavior. A formal definition of a self-similar fractal in a two-dimensional x-y-space is that $f(rx, ry)$ is statistically similar to $f(x, y)$ where r is a scaling factor. This may be quantified by applications of the fractal relation

$$N = C r^{-D} \tag{5.1}$$

where r is a characteristic linear dimension, D is the fractal dimension (real number >0), C is a constant of proportionality, the pre-factor parameter, $N=N(>r)$ is the number of objects with characteristic linear dimension $\geq r$.

As example, the number of boxes with dimension x_1 and y_1 required to cover a given object is N_1 and the number of boxes with dimensions $x_2 = r x_1$, $y_2 = r y_1$ required to cover the object is N_2. If the object is a self-similar fractal, we have that

$$N_2/N_1 = r^{-D}$$

In the same manner one may consider a self-similar fractal in a n-dimensional $x_1, x_2, \dots\dots, x_n$-space with $f(rx_1, rx_2, \dots\dots, rx_n)$ statistically similar to $f(x_1, x_2, \dots\dots, x_n)$. with r scaling factor.

Many physiological processes posses scale similarity (scale-invariance) properties. Self-similarity or statistical self-similarity may be investigated in given time series data with the finality to establish their fractal or multi fractal behavior (16). In the present paper we will adopt the following simple procedure.

Let us take now the notion of variogram previously exposed.

Consider the importance to have introduced here an analysis by variogram of time series regarding state anxiety. While the previously introduced indexes SD1 and SD2 give a general indication on variability in time of state anxiety in short as well as in long time intervals, variogram enables to quantify such time variability at each lag time. In particular, a small value of the variogram will indicate that pairs of results of the given time series are similar or have a low variability at a particular time distance of separation. Of course, high values of the variogram will indicate instead that the values are very dissimilar or that we have high variability.
The results of the variograms are reported in Figures 8-13 while the results of the fractal analysis are reported in Table 8.

The analysis of the variograms reveals some important features. The subjects F. Dav and A. Men gave the most modest values of variograms ranging from 0 to 8.34 at least. They had a low

Journal of Consciousness Exploration & Research| December 2010 | Vol. 1 | Issue 9 | pp. 1-69   34
Conte, E., Todarello, O., Conte, S., Mendolicchio, L., Mendolicchio, L. & Federici, A.
Methods and Applications of Non-Linear Analysis in Neurology and Psycho-physiology

variability in time. This result is in accord with the mean value of the test that in fact gave the lowest values for such two subjects. Also the statistical values of st. dev. and of variance resulted very contained. Corresponding such subjects gave also the lowest values for SD1 and SD2 respectively. Note also that the variogram showed the tendency to decrease progressively (decreasing variability) after 20 lags (about 5 days) and to annul itself in about 80-90 lags corresponding to about 500 hours. This behavior reveals the tendency of state anxiety in such two subjects to vary with some periodicity concluding its cycle in about twenty days. Of course the tendency of the variogram to a progressive decrease (progressively decreasing variability) resulted mixed to time lag intervals with variogram showing increased variability as example, at 30, 50, 70 time lags corresponding to 180, 300, and 420 hours.

Soon after, the subject A. Mac showed a more marked variability with a variogram ranging from 0 to 43.51. It is important to outline that also in this case we have an excellent agreement with the mean value of the test, the statistical indexes and the values of SD1 and SD2 respectively. Also in this case the variogram showed the tendency to decrease progressively its variability and to annul itself in about 80 lags. Again it followed an initial increase until 20 lags and still we had mixed time lag intervals of increasing variability at about 30, 50, 70 time lags.

The same important results are obtained by inspection of the variogram regarding the subject G. Den. In this case the score of the test was of 53.4 in mean with a st. dev. of 8.8 and variance of 82.67. Correspondingly, the variogram also increased its maximum value ranging this time from 0 to 84.95. Also Sd1 and SD2 increased their values. The behavior of the variogram remained the same as in the previous cases, differing only for the assumed values. In particular, it increased until a time lag of about 20 lags and thus it decreased progressively and annulled itself in about 80-90 lags with mixed peaks at about 30, 50, 70 lags. In substance, they were the values of the variogram to differentiate the behavior of this subject respect to the other subjects while apparently the time dynamics remained unvaried for all the examined subjects. The same conclusions may be reached examining the case of the subject M. Den. This time the mean value of the test reached 47.2 with a st. dev. of 10.3 and a variance of 119.77. Respect to the subject G. Den, the score of the test resulted in mean lightly less (47.2 vs 53.4) but the st. dev. (10.3 vs 8.8) and the variances resulted greater (119.77 vs 82.67). The corresponding variogram assumed still an higher value ranging this time from 0 to 130.9. In correspondence also SD1 and SD2 resulted strongly increased and, in detail, SD1 resulted to be 4.72 and SD2 assumed the value of 10.3.

The time lag behavior resulted the same as in the previous cases with the exception, obviously, of the assumed values. Also this time it increased until about 20 lags and thus it started to decrease annulling itself about 80-90 lags. Peaks were found again about 30, 50, 70 lags.

The subject D. Pet showed instead some important differences. He had a mean score of the test of 38.3 (very similar to the score 38.1 of the subject A. Mac), but he had a st.dev. of 9.00 and a variance of 99.89, an high value in the investigated group. In correspondence SD1 resulted to be 58.2 but SD2 resulted to be 12.96, the highest value in the group of subjects. In conclusion, in spite of a moderate value of his score (38.3 mean value), this subject showed the highest variability in time dynamics of his state anxiety. In correspondence, the variogram resulted ranging from 0 to 100 and the time lag dynamics showed some modifications respect to the case of the other subjects. As in the previous cases, it increased until 20 lags and than it started to

Journal of Consciousness Exploration & Research| December 2010 | Vol. 1 | Issue 9 | pp. 1-69   35
Conte, E., Todarello, O., Conte, S., Mendolicchio, L., Mendolicchio, L. & Federici, A.
Methods and Applications of Non-Linear Analysis in Neurology and Psycho-physiology

decrease but annulling itself, this time about 110-120 lags. Still, very marked peaks this time appeared at about 20, 40, 60, 80-90 lags. In brief, we had some modifications in time dynamics but especially in the time variability of the dynamics of this subject.

The explanation may be found analyzing the results of fractal analysis that are given in Table 8. Before of all, we have to outline that our analysis indicates for the first time that state anxiety responds to a fractal structure. We have a kind of multiplicative process possibly supported by additive noise. The Fractal Measure more than the Generalized Fractal Dimension reveals that we had moderate values of such parameter in correspondence of low values for mean score of the test, of st.dev., of variance, of SD1 and SD2 while instead we had progressively increasing values of Fractal Measure for increasing values of the mean value of score, of st.dev., of variance and of SD1 and SD2 in the case of the other subjects. In spite of a rather stable value for Generalized Fractal Dimension, we had value for Fractal Measure that progressively range from 13.2 to 305.00 with a net differentiation and thus a discriminating ability.

5.6 Linear Analysis in Frequency Domain

In order to deepen the results that we obtained about the recurrent components that we identified by variogram analysis, we performed a further preliminary analysis calculating Fourier spectrum of the time series data of the six examined subjects. We must remember here that the limit of this kind of analysis is that it is a linear method in the framework of a process that instead is intrinsically non linear. However, we arrived to obtain some preliminary interesting information. In frequency domain we calculated an AR spectrum by using an AR model at order 16. We evidence for the first time that time behavior of state anxiety exhibits some harmonic components peaked at some specific frequencies that we identified in all the examined subjects. The basic features of such spectra are summarized in Figures 14-19 and in Table 9. We identified four bands of interest. The first about 0.1 Hz, the second about 0.2 Hz, the third in the region 0.3-0.4 Hz, the fourth about 0.5 Hz. Note that in ordinate we have always the test score as reference. The actual value is obtained by square root of power spectrum and multiplying by 100.

As we know, we sampled the time series of subjects at time steps of about three hours. The frequency value was of 9.25×10^{-5} Hz. The spectra are given, in accordance with the Nyquist theorem, at 0.5 of such value. By such analysis it is seen that four peaks are always present in all the examined spectra. All they are given at the following times. One period of time is about 30 hours, the other is given about 15 hours, the third about 7.5-10 hours and, finally, the last about 6 hours. We find for the first time that state anxiety runs again quite periodically with times of 30,15,7.5-10, 6.0 hours. This is a very important result also for diagnostic and therapeutic reasons.

Conclusions

We started admitting a quantum like model for behavior of mind entities in state anxiety of human subjects. Our aim was to investigate time dynamic variability of data preparing several experimental time series that were obtained by using the well known test of D. Spielberger as it was arranged starting with 1964. We obtained that the dynamic of this process follows a fractal

Journal of Consciousness Exploration & Research| December 2010 | Vol. 1 | Issue 9 | pp. 1-69   36
Conte, E., Todarello, O., Conte, S., Mendolicchio, L., Mendolicchio, L. & Federici, A.
Methods and Applications of Non-Linear Analysis in Neurology and Psycho-physiology

regime possibly a quantum fractal behavior (18). In order to proceed with a quantification of the basic features of the time series under investigation, in addition to fractal analysis, we introduced several parameters, and, in detail, the Poincaré-plots with linked the indexes SD1 and SD2, quantifying time variability of the data along short as well as long times, and an analysis of time series data by a variograms. We found that SD1 and SD2 are very satisfactory indexes that may be used to characterize in detail time variability of state anxiety in human subjects. Also analysis by variograms confirmed its predictive attitude. It resulted able to delineate time variability of state anxiety at each time step and to differentiate among the different conditions of variability of human subjects. In particular, the use of variogram analysis enabled us also to identify an important feature of dynamics of the engaged mind entities. We found that the variograms of the different six subjects exhibit some constant recurrences in lags and thus in time: the variograms assumed always the same increasing and decreasing behavior at about the same times with pronounced peaks of variability still at the same recurrent times and finally such variograms annulling themselves also at recurrent times. This result suggests that the engaged mind entities behave in time following a proper inner function. In fact, the variograms of different subjects presented the same kinds of recurrences in time for all the subjects, also submitted to different environmental conditions. In conclusion, the state anxiety seems to represent an emotional human condition that is so general and so radical in human to express a common mode of human existence in time, regulated in the inner of mind entities by the same recurrent, deterministic like, function. In particular it was estimated by us that such recurrent mind function seems to repeat itself with periodicity like of about twenty days and giving again basic features of self-similarity. This recurrent function results instead to be differentiated in subjects, from subject to subject, only for the different values that it assumes at the same prefixed times. In conclusion, the state anxiety shows a rather constant tendency to be recurrent in time with an inner deterministic - periodic like mechanism. Harmonic components were also found when we submitted time series data to frequency domain analysis by FFT.

Finally, there are some other important questions that we examined in the present paper. We attributed a great importance to the analysis of the time variability of the data of time series that were investigated.. The different scores that were obtained in mean for the test of the six subjects, linked to the different values of st.dev. and variances, SD1 and SD2, and compared with the results of fractal analysis, indicated that the increasing mean values of the score of the test, of st. dev., of variance and of SD1 and SD2 correspond to an increasing time variability of the data in a recurrent functional framework that of course remained instead rather constant for all the six subjects not in the assumed values but in the temporal behavior. As general representation of the process, it seems thus emerging a framework in which we have a basic recurrent, deterministic like, process whose time behavior remains rather similar for all the six subjects but it is, instead, differentiated from subjects to subject owing to the variations and variability in the values that in time such basic function assumes in correspondence of the different states of anxiety characterizing the different subjects. This seems to represent an interesting result that we would comment in more detail.

As previously we said, state anxiety represents an emotional condition that is so general and so radical in human that it cannot be considered only a sign of pathology or a defined syndrome but a general mode of the human existence. In our opinion, it represents consequently a proper condition to investigate on a general plane some features of mind entities in analysis of their time

Journal of Consciousness Exploration & Research| December 2010 | Vol. 1 | Issue 9 | pp. 1-69   37
Conte, E., Todarello, O., Conte, S., Mendolicchio, L., Mendolicchio, L. & Federici, A.
Methods and Applications of Non-Linear Analysis in Neurology and Psycho-physiology

dynamic pattern. In state anxiety, anxiety is activated uniquely in the conditions in which the subject evaluates his living situations as a threat and consequently he activates a sequence of behaviors as generally they are induced from anxiety. We have suggested a quantum like model for this process assuming a superposition of potential states in mind entities before the subject actives introspection. In detail our model runs as it follows.

1- State anxiety rises on the basis of inner motivations of the subject.

This is to say that an inner stimulus as thoughts, feelings, biological needs,…is configured in mind entities as a superposition of potentialities. To give an example, remaining on the general plane let us examine the kind of emotional response that a subject could give to an event. In the case of a quantum like superposition of potentialities , we will have the following indicative expression

$$\psi = c_1|frustrated > + c_2|anxious > + c_3|excited > + c_4|angry > + \ldots\ldots \qquad (5.2)$$

where ψ will represent the whole potential state of the mind entity of the subject for the emotion response . Each $|\ldots >$ will represent each potential state of the emotion response dynamics and the complex numbers c_i (i = 1,2,…..) will be probability amplitudes so that $|c_i|^2$ will represent the probability that the potential state i will be actually recognized (actualized) at cognitive level when the subject will actualize his response thinking about his situation.

2- At the level of state anxiety, the initial stimulus will be inner (thoughts, feeling, …..) and, still again, we will have a superposition of potentialities as response. As example, with regard to the possibility for the subject to feel excitement as consequence of such inner stimulus, we will have
$$\psi = c_1| very\ moderate\ excitement >+c_2| moderate\ excitement >+ c_3| quite\ high\ excitement >+c_4| very\ high\ excitement > \qquad (5.3)$$

This is the superposition of potentialities at the level of mind entities.

3-The following step is that the subject will perform a cognitive evaluation. He will perform an introspective activity, an act of consciousness, and by this act, he will give actualization to only one among the various potentialities before mentioned in the (5.2 or 5.3). He will perform a transition potentiality → actualization giving to himself to be in the actual state 1 or 2 or 3 or 4. The first actualization will be performed with probability $|c_1|^2$, the second with probability $|c_2|^2$, the third with probability $|c_3|^2$ and the fourth with probability $|c_4|^2$. One actualization among the different possibilities will give also a score as result of the answer given from the subject submitted to the various questions posed by the test. The same mechanism will happen for the other posed questions of the test.

Note that the particular importance of the (5.2 or 5.3) resides in the term superposition that we have employed for it. The potential states (5.2 or 5.3) represents the simultaneous presence of the four potentialities in mind entities of the subject. Consequently, the deriving model is that one of an intrinsic indetermination for mind entity at this stage. Such intrinsic and ontological indetermination is released only at the moment of the individual cognitive evaluation when he actualizes one and only one of the possibilities at his disposal and cancels the previous

Journal of Consciousness Exploration & Research| December 2010 | Vol. 1 | Issue 9 | pp. 1-69   38
Conte, E., Todarello, O., Conte, S., Mendolicchio, L., Mendolicchio, L. & Federici, A.
Methods and Applications of Non-Linear Analysis in Neurology and Psycho-physiology

indetermination.

4- As consequence of the actualization during the cognitive evaluation, the subject will experience a number of subjective feelings, of apprehensions, of anxious expectations also with activation (arousal) of his nervous system, and with the final evidence of some subjective behaviors.

5-Some control and /or defence mechanisms will interfere with this dynamic. They will have the finality to give adaptability to the subject and reduction of anxiety.

The time series data that we collected for the six subjects reflect in some manner all this time dynamics, and we must expect that, in correspondence to the different mean values that were obtained as result of the test, the different subjects characterized the different levels of indetermination that, as previously seen, represent the crucial point of the whole process generating state anxiety.

Let us give still some examples in order to be clear. For a subject with a very moderate or moderate mean value of the test of state anxiety we should have that the values of probabilities $|c_1|^2$ and $|c_2|^2$ of the (5.2 or 5.3), just corresponding to a moderate anxiety, will be very high while there will be present very low values of probabilities of $|c_3|^2$ and $|c_4|^2$, corresponding instead to high anxiety. We will have approximately that

$$|c_1|^2 + |c_2|^2 \equiv 1 \qquad \text{and} \qquad |c_3|^2 \to 0 \text{ and } \quad |c_4|^2 \to 0 \qquad\qquad (5.4)$$

This will be true for all the questions posed to the subject during the test.

A subject having a very moderate or a moderate anxiety will be suspended really between two potential states (1) and (2) instead of (1), (2), (3), (4), being $|c_3|^2 \equiv 0$ and $|c_4|^2 \equiv 0$ and thus he will have a more moderate indetermination respect to the general case.

The subjects with an high mean value of the score and an high variability in time, will have

$$|c_1|^2 + |c_2|^2 + |c_3|^2 + |c_4|^2 \equiv 1 \quad \text{with } |c_3|^2 + |c_4|^2 > |c_1|^2 + |c_2|^2 \qquad\qquad (5.5)$$

with all the four potential states having the concrete possibility of being actualized and thus such subjects will show greater indetermination and greater variability of data respect to the previous case.

We may say that in the first case we have a lower indetermination in the potential states respect to the second one. This is to say that the subjects having higher mean score should exhibit more elevate indetermination respect to the case of subjects with less mean score. Obviously, in the case of more elevate values for scores and thus of indetermination, we expect that more hardly control mechanisms acted to reduce state anxiety or to induce adaptability in the investigated subjects, and such systematic action of mind and biological control induced high variability in the measured data. This is the reason because we found so marked differences in st. dev., in variances, in SD1 and SD2, and in variograms in the different examined subjects. The found time

Journal of Consciousness Exploration & Research| December 2010 | Vol. 1 | Issue 9 | pp. 1-69
Conte, E., Todarello, O., Conte, S., Mendolicchio, L., Mendolicchio, L. & Federici, A.
Methods and Applications of Non-Linear Analysis in Neurology and Psycho-physiology

39

variability of data was also direct expressions of the acting mechanism of mental and biological control and defence, and this was, in conclusion, the reason because, from its starting, we attributed so much attention to our analysis of time variability of data. They are indications of the great indetermination that is at the basis of this process as well as of the basic mechanisms of control that consequently enter in action. They, of course, represent the central core of the mechanisms to be understood in analysis of state anxiety. It is this reason because the quantitative indexes, that we introduced, seem to be of relevant importance. They are just able to characterize and to quantify indeterminism and acting control mechanisms in the dynamics of state anxiety of subjects.

References

[1] Zak M., The problem of irreversibility in newtonian dynamics. International Journal of Theoretical Physics 1992; 31 (2): 333-342.

Zak M., Non-Lipschitzian dynamics for neural net modelling. Appl. Math. Lett. 1989; 2 (1): 69-74.

Zak M., Non-Lipschitz approach to quantum mechanics. Chaos, Solitons and Fractals 1998; 9 (7): 1183-1198.

Zbilut J.P., Hubler A, Webber Jr. CL., Physiological singularities modeled by nondeterministic equations of motion and the effect of noise. In Milonas, M (Ed) Fluctuations and Order: The New Synthesis. New York: Springer Verlag. 1996: 397-417.

Zbilut J.P., Zak M, Webber Jr. CL., Nondeterministic chaos in physiological systems. Chaos, Solutions, and Fractals 1995; 5: 1509-1516.

Zbilut J.P., Zak M, Webber, Jr. CL., Nondeterministic chaos approach to neural intelligence. Intelligent Engineering Systems Through Artificial Neural Networks. Vol. 4. New York: ASME Press, 1994: 819-824.

[2] Conte E., Federici A., Zbilut J.P., On a simple case of possible non-deterministic chaotic behaviour in compartment theory of biological observables. Chaos, Solitons and Fractals 2004; 22 (2): 277-284

Conte E., Pierri GP., Federici A., Mendolicchio L., Zbilut J.P., A model of biological neuron with terminal chaos and quantum-like features. Chaos, Solitons and Fractals 2006; 30 (4): 774-780

Conte E., Vena A., Federici A., Giuliani R., Zbilut J.P., A brief note on possible detection of physiological singularities in respiratory dynamics by recurrence quantification analysis of lung sounds. Chaos, Solitons and Fractals 2004; 21 (4): 869-877.

Vena A., Conte E., Perchiazzi G., Federici A., Giuliani R., Zbilut J.P., Detection of physiological singularities in respiratory dynamics analyzed by recurrence quantification analysis of tracheal sounds. Chaos, Solitons and Fractals 2004; 22 (4): 869-881.

[3] Webber Jr. CL., Zbilut J.P., RQA of nonlinear dynamical systems. www.nsf.gov/sbe/bcs/pac/nmbs/chap2.pdf

[4] Takens F., Detecting strange attractors in turbulence. Lectures notes in mathematics 1981; 898, Takens F., Dynamical systems and turbulence, Warwick, Berlin, Springer-Verlag,1980: 336-381.

For an excellent exposition of concepts of phase space and, in general, of methodologies and conceptual foundations in fractal and chaos theory, see the book:

Journal of Consciousness Exploration & Research| December 2010 | Vol. 1 | Issue 9 | pp. 1-69   40
Conte, E., Todarello, O., Conte, S., Mendolicchio, L., Mendolicchio, L. & Federici, A.
Methods and Applications of Non-Linear Analysis in Neurology and Psycho-physiology

Bassingthwaighte J.B, Liebovitch L., West B. J., Fractal Physiology. published by the American Physiological Society, New York, Oxford, 1994.

[5] Fraser AM., Swinney HL., Independent coordinates for strange attractors from mutual information. Phys. Rev. 1986; A 33: 1134-1140.

[6] Kennel MB., Brown R., Abarbanel HDI., Determining embedding dimension for phase space reconstruction using a geometrical construction, Pys. Rev. A 1992; 45: 3403-3411.

[7] Grassberger P., Procaccia I., Measuring the strangeness of strange attractors. Physica D 1983; 9: 189-208.

[8] Eckman J.P., Ruelle D., Ergodic theory of chaos and strange attractors. Rev. Mod. Phys. 1986; 57: 617-656.

[9] Holden A.V., Chaos, Manchenster University Press, Manchester, 1986.

Holzfuss J., Mayer–Kress G., Dimensions and entropies in Chaotic systems, quantification of complex behaviour. Spriger Verlag, Berlin, Heidelberg, 1986.

Pesin Y.B., Characteristic Lyapunov Exponents and Smooth Ergodic Theory. Russian Math. Surveys 1977, 32 (4): 55-114.

[10] Christiansen F., Rugh H.H., Computing Lyapunov spectra with continuous Gram-Schmidt orthonormalization. Nonlinearity 1997; 10: 1063–1072.

Habib S. and Ryne R.D., Symplectic Calculation of Lyapunov Exponents. Physical Review Letters 1995; 74: 70–73.

Rangarajan G., Habib S, Ryne R.D., Lyapunov Exponents without Rescaling and Reorthogonalization. Physical Review Letters 1998; 80: 3747–3750.

Zeng X., Eykholt R., Pielke R.A., Estimating the Lyapunov-exponent spectrum from short time series of low precision. Physical Review Letters 1991; 66: 3229.

Aurell E., Moffetta G., Frisanti A., Paladin G., Vulpiani A, Predictability in the large: an extension of the concept of Lyapunov exponent. J. Phys. A: Math. Gen. 1997; 30: 1–26.

[11] Mandelbrot B.B., Van Ness J.W., Fractional Brownian motions, fractional noises and applications, SIAM Rev 1968; 10: 422-436.

Mandelbrot B.B, The fractal geometry of Nature. Freeman, San Francisco, 1982.

[12] Eckmann J.P., Kamphorst S.O., Ruelle D., Recurrence Plots of dynamical systems. Europhysics Letters 1987; 4: 973-977.

[13] Webber C.L., Zbilut J.P., Dynamical assessment of physiological systems and states using recurrence plot strategies. Journ. of Applied Physiology 1994; 76: 965-973.

[14] Marwan N., Encounters with neighbors: current developments of concepts based on recurrence plots and their applications, doctoral dissertation. Institute of Physics, University of Postdam, 2003.

Marwan N., Thiel M., Nowaczyk N.R., Cross recurrence plot based synchronization of time series. Nonlinear processes in geophysics 2002; 9: 325-331.

[15] Conte E., Federici A., Minervini M., Papagni A., Zbilut J.P, Measures of coupling strength and synchronization in non linear interaction of heart rate and systolic blood pressure in the cardiovascular control system. Chaos and Complexity Letters 2006; 2 (1): 1-22.

[16] Conte E., Federici A., Pierri GP, Mendolicchio L., Zbilut J.P., A Brief Note on Recurrence Quantification Analysis of Bipolar Disorder Performed by Using a van der Pol Oscillator. Chaos and Complexity Letters 2007; 3 (1): 25-44.

Orsucci F., Nonlinear dynamics in language and psychobiological interactions. in 'Orsucci, F. (ed) The Complex Matters of the Mind, World Scientific, Singapore and London 1998.

Orsucci F., Walters K., Giuliani A., Webber Jr CL., Zbilut J.P., Orthographic Structuring of

Journal of Consciousness Exploration & Research| December 2010 | Vol. 1 | Issue 9 | pp. 1-69   41
Conte, E., Todarello, O., Conte, S., Mendolicchio, L., Mendolicchio, L. & Federici, A.
Methods and Applications of Non-Linear Analysis in Neurology and Psycho-physiology

Human Speech and Texts: Linguistic Application of Recurrence Quantification Analysis, International Journal of Chaos Theory and Applications 1999; 4 (2): 80–88.

Orsucci F., Changing Mind: transitions in natural and artificial environments. World Scientific, Singapore and London 2002.

Orsucci F., Giuliani A, Webber CL. Jr, Zbilut J.P., Fonagy P., Mazza M., Combinatorics and synchronization in natural semiotics, Physica A: Statistical Mechanics 2006; 361: 665

[17] Mastrolonardo M., Conte E., Zbilut J.P., A fractal analysis of skin pigmented lesions using the novel tool of the variogram technique. Chaos, Solitons and Fractals 2006; 28 (5): 1119-1135.

Conte E., Pierri GP., Federici A., Mendolicchio L., Zbilut J.P., The Fractal Variogram Analysis as General Tool to Measure Normal and Altered Metabolism States and the Genetic Instability: An Application to the Case of the Cutaneous Malignant Melanoma. Chaos and Complexity Letters 2008; 3 (3): 121-135.

Conte E., Federici A., Zbilut J.P., A new method based on fractal variance function for analysis and quantification of sympathetic and vagal activity in variability of R–R time series in ECG signals. Chaos, Solitons and Fractals 2008; 41 (2009) 1416–1426

Conte E., Khrennikov A., Federici A., Zbilut J.P., Fractal Fluctuations and Quantum-Like Chaos in the Brain by Analysis of Variability of Brain Waves: A New Method Based on a Fractal Variance Function and Random Matrix Theory. arXiv:0711.0937,

Conte E., Khrennikov A., Federici A., Zbilut J.P., Fractal Fluctuations and Quantum-Like Chaos in the Brain by Analysis of Variability of Brain Waves: A New Method Based on a Fractal Variance Function and Random Matrix Theory: a link with El Naschie fractal Cantorian space time and H. Weiss and V. Weiss golden ratio in brain.. Chaos, Solitons and Fractals. 41 (2009) 2790–2800

[18] Hurst H.E., Black R., Sinaika Y.M., Long term storage in reservoirs: an experimental study. Costable, London, 1965.

[19] Wei S. and Pengda Z., Multidimensional Self-Affine Distribution with Application in Geochemistry. Mathematical Geosciences 2002; 34 (2): 121-131.

[20] http://physionet.org/physiobank/database/mvtdb/RRdata1/
http://www.physionet.org/physiobank/database/fantasia/subset/

[21] Giuliani A., Piccirillo G., Marigliano V., Colosimo A, A nonlinear explanation of aging-induced changes in heartbeat dynamics. Am J Physiol Heart Circ Physiol 1998; 275: H1455-H1461.

[22] El Nashie M.S., E–finite theory – some recent results and new interpretations. Chaos, Solitons and Fractals 2006; 29: 845-853.

[23] Weiss H. and Weiss V., The golden mean as clock cycle of brain waves. Chaos, Solitons and Fractals 2003; 18: 643-652.
See also Datta D.P. and Raut S., The arrow of time and the scale free analysis. Chaos, Solitons and Fractals 2006; 28: 581-589.

[24] Koelsch S., Remppis A., Sammler D., Jentschke S., Mietchen D., Fritz T., Bonnemeier H. Walter A. Siebel5 W. A cardiac signature of emotionality, European Journal of Neuroscience, 2007, 26: 3328–3338,

[25] Thayer J.F., Smith M., Rossy L.A., Sollers J., Friedman B.H. Heart Period Variability and Depressive Symptoms: Gender Differences, Biol. Psychiatry, 1998, 44: 304–306

Journal of Consciousness Exploration & Research| December 2010 | Vol. 1 | Issue 9 | pp. 1-69
Conte, E., Todarello, O., Conte, S., Mendolicchio, L., Mendolicchio, L. & Federici, A.
Methods and Applications of Non-Linear Analysis in Neurology and Psycho-physiology

42

References for analysis of state anxiety

[1] Larsen R. J., 1989, A process approach to personality psychology, D.M. Buss and N. Cantor Eds, Personality and Psychology, Recent Trends and Emerging Directions, Springer Verlag, New York.

[2] Totterdell P., Briner R.B., Parkinson B., Reynolds S., Fingerprinting Time Series: Dynamic patterns in self-report and performance measures uncovered by a graphical non linear method, British Journal of Psychology, 1996, 87: 43-60

[3] Tennen H., Suls J., Affleck G., Personality and daily experience: The promise and the challenge, Journal of Personality, 1991: 59: 313-338

[4] DeLongis A., Folkman S., Lazarus R.S., The impact of daily stress on health and mood: psychological and social resources as mediators, Journal of Personality and Social Psychology, 1988, 54: 486-495

[5] Brandstatter H., 1991, Emotions in every day life situations.Time sampling of subjective experience, Subjective well being, Oxford, Pergamon Press.

[6] Larsen R. J, Kasimatis M., Day to day physical symptoms:individual differences in the occurrence, duration and emotional concomitans of minor daily illness, Journal of Personality, 1991, 59: 387-424

[7] Zak M., Non-Lipschitz approach to quantum mechanics, Chaos Solitons and Fractals, 1998, 9 (7): 1183-1198 and references therein.

Zbilut J.P., 1997, From Instability to Intelligence: Complexity and Predictability in nonlinear Dynamics, Lecture Notes in Physics, New Series m 49, Springer Verlag.

Zbilut J.P., 2004, Unstable singularities and Randomness: Their importance in the complexity of physical, biological and social sciences, Elsevier Science.

Gabora L. Aerts D. Creative Thought as a non Darvinian evolutionary process, Journal of Creative Behavior (in press), and references therein.

Aerts D, Broekaert J., Gabora L., A case for applying an abstracted quantum formalism to cognition, M.H. Bihard and R.Campbell Eds, Mind in interaction, Amsterdam: John Benjamin Archive (in press), quant-ph/0404068, and references therein.

Conte E., Pierri G.P., Federici A., Mendolicchio L., Zbilut J.P., On a model of biological neuron with terminal chaos and quantum like features, Chaos, Solitons and Fractals, Chaos Solitons and Fractals, 2006, 30: 774-780.

[8] Conte E., Todarello O., Federici A., Vitiello F., Lopane M., Khrennikov A, Zbilut J.P., Found Experimental Evidence of Quantum Like Behavior of Cognitive Entities. An abstract quantum mechanical formalism to describe cognitive entity and its dynamics, Chaos, Solitons and Fractals, Chaos Solitons Fractals 2006, 31: 1076-1088.

Conte E., Todarello O., Federici A., Vitiello F., Lopane M., Khrennikov, 2003, A preliminar evidence of quantum like behavior in measurements of mental states, Quantum Theory: Reconsideration of Foundations, Ed. A.Yu. Khrennikov, Ser. Math. Modeling in Phys. Eng. and Cognitive Sciences, vol.3, Vaxjo, Univ. Press., and references therein;

Conte E,.Khrennikov A., Todarello O., Federici A., Mendolicchio L., Zbilut J.P , Mental States Follow Quantum Mechanics During Perception and Cognition of Ambiguous Figures, Open Systems and Information Dynamics 2009, 16 (1): 1-17.

Conte E., Khrennikov A., Todarello O., Federici A., Zbilut J.P, On the Existence of Quantum Wave Function and Quantum Interference Effects in Mental States An Experimental Confirmation during Perception and cognition in humans, Neuroquantology, June 2009, 7 (2): 204-212 and references therein.

Journal of Consciousness Exploration & Research| December 2010 | Vol. 1 | Issue 9 | pp. 1-69
Conte, E., Todarello, O., Conte, S., Mendolicchio, L., Mendolicchio, L. & Federici, A.
Methods and Applications of Non-Linear Analysis in Neurology and Psycho-physiology

43

Conte E. Exploration of Biological function by quantum mechanics, Proceedings 10th International Congress on Cybernetics, 1983;16-23, Namur, Belgique.

Conte E. Testing Quantum Consciousness NeuroQuantology 2008; 6 (2): 126-139

Elio Conte, A Proof That Quantum Interference Arises in a Clifford Algebraic Formulation of Quantum Mechanics , available on line Philpapers

Elio Conte, On Some Cognitive Features of Clifford Algebraic Quantum Mechanics and the Origin of Indeterminism in This Theory: A Derivation of Heisenberg Uncertainty Principle by Using the Clifford Algebra, available on line PhilPapers

Elio Conte (2009). Decision Making : A Quantum Mechanical Analysis Based On Time Evolution of Quantum Wave Function and of Quantum Probabilities During Perception and Cognition of Human Subjects, available on line Philpapers.

Conte E., A reformulation of von Neumann's postulates on quantum measurement by using two theorems in Clifford algebra, International Journal of Theoretical Physics, DOI 10.1007/s10773-009-0239-z , available on line

Khrennikov A. Yu., Linear Representation of probabilistic transformations induced by context transitions, J, Phys. A. Math.and Gen., 2001, 34: 9965-9981;

Khrennikov A. Yu, Representation of the Kolmogorov model having all distinguishing features of quantum probabilistic model, Phys. Lett. A., 2003, 316: 279-296.

[9] Catherall Don R., How Fear differs from anxiety, Traumatology, 2003, 9 (2): 76-92.

[10] Lang P.J. Davis A., Fear and Anxiety: animal models and human cognitive psycophysiology, Journal of affective disorders, 2000, 61 (3): 137-159.

[11] Sullivan G.M., Copland J.D., Kent J.M., Gorman J.M., The adrenergic system in pathological anxiety: a focus on panic with relevance to generalized anxiety and phobias, Biological Psychiatry, 1999, 46 (9): 1205-1218

[12] Spielberger C.D. 1991, S.T.A.I., Ansia di Stato e di Tratto, Wyeth, Organizzazioni Speciali, Firenze.

[13] Brenman M., Palaniswami M., Kamen P., (2001), Do existing measures of Poincaré plot geometry reflect non linear features of HRV?, IEE transactions on biomedical Eng., and references therein

[14] Sprott J.C., 2003, Chaos and Time series analysis, Oxford University Press.

[15] Mandelbrot B.B., 1975, Les objects fractals: forme, hasard et dimension, Paris Flammarion

[16] Merlini D., Losa G., 2005, Fractals in Biology and Medicine, Springer Verlag.

[17] Wei S., Pengda Z., Multidimensional self- affine distributions with application in geochemistry, Math. Geol., 2002, 34 (2):109-123, and references therein.

[18] see as example:

Gomez J.M.G., Relano A., Retamosa J., Failero E., Salasnich L., Vranicar M., Robnik M., 1/f Noise in Spectral Fluctuations of quantum systems, Physical Review Letters, 2005, 94: 84101-84104;

Relano A., Gomez J.M.G., Molina R.A., Retanosa J., Quantum Chaos and 1/f noise, Phys. Review Letters, 2002, 89: 24102-24105.

[19] Webber C.L. Jr, Zbilut J.P. Dynamical assessment of physiological systems and states using recurrence plot strategies, J. Appl. Physiol. 1994, 76: 965-973.

Journal of Consciousness Exploration & Research| December 2010 | Vol. 1 | Issue 9 | pp. 1-69
Conte, E., Todarello, O., Conte, S., Mendolicchio, L., Mendolicchio, L. & Federici, A.
Methods and Applications of Non-Linear Analysis in Neurology and Psycho-physiology

44

Table 1 Embedding Analysis

Subjects	Autocorrelation Function (Au)	Mutual Information (MI)	False Nearest Neighbors (FNN)
Y_1	10	3	5
Y_2	103	1	5
Y_3	32	3	5
Y_4	17	2	5
Y_5	19	2	5
O_1	12	1	4
O_2	385	2	4
O_3	110	2	4
O_5	16	3	4
O_6	23	3	4
Vt_3-13	312	2	4
Vt_2-67	86	4	4
Vt_1-26	32	3	7
Vt_1-15	357	2	2
Vt_1-03	139	2	6
Vf_2-30	21	4	8
Vf_2-71	1	4	5
Vf_1-8013	90	4	4
Vf_1-217	1	2	4
Vf_1-115	358	3	2

Table 2 Largest Lyapunov Exponent

Subjects	λ_E		Statistical analysis (t-Test)	
Y_1	0.625 ± 0.054		Y_i vs O_i	
Y_2	0.635 ± 0.052		P value	0.0066
Y_3	0.645 ± 0.055		P value summary	**
Y_4	0.625 ± 0.049		Are means signif. different? (P < 0.05)	Yes
Y_5	0.521 ± 0.053		t, df	t=3.634 df=8
O_1	0.562 ± 0.047		O_i vs Vt_i	
O_2	$0.440 \pm .044$		P value	0.0281
O_3	0.523 ± 0.052		P value summary	*

Journal of Consciousness Exploration & Research| December 2010 | Vol. 1 | Issue 9 | pp. 1-69
Conte, E., Todarello, O., Conte, S., Mendolicchio, L., Mendolicchio, L. & Federici, A.
Methods and Applications of Non-Linear Analysis in Neurology and Psycho-physiology

45

O_5	0.439 ± 0.055	Are means signif. different? (P < 0.05)	Yes
O_6	$0.490 \pm .066$	t, df	t=2.675 df=8
Vt_3-13	0.373 ± 0.063	**O_i vs Vf_i**	
Vt_2-67	0.432 ± 0.085	P value	0.787
Vt_1-26	0.430 ± 0.094	P value summary	ns
Vt_1-15	0.150 ± 0.058	Are means signif. different? (P < 0.05)	No
Vt_1-03	0.294 ± 0.113	t, df	t=0.2794 df=8
Vf_2-30	0.498 ± 0.122	**Y_i vs Vt_i**	
Vf_2-71	0.605 ± 0.083	P value	0.0014
Vf_1-8013	0.648 ± 0.074	P value summary	**
Vf_1-217	0.668 ± 0.066	Are means signif. different? (P < 0.05)	Yes
Vf_1-115	0.168 ± 0.098	t, df	t=4.777 df=8
		Y_i vs Vf_i	
		P value	0.3567
		P value summary	ns
		Are means signif. different? (P < 0.05)	No
		t, df	t=0.9780 df=8
		Vt_i vs Vf_i	
		P value	0.1257
		P value summary	ns
		Are means signif. different? (P < 0.05)	No
		t, df	t=1.710 df=8

Table 3 RQA Analysis

Subjects	% Rec	% Det	% Lam	T.T.	Ratio	Entropy	Max Line	Trend
Y_1	0.171	0.342	0.685	3.000	2.000	0.000	3	0.090
Y_2	0.391	33.842	0.148	3.000	86.638	1.491	8	-0.252
Y_3	0.132	1.037	0.000	0.000	7.859	0.000	7	-0.216
Y_4	0.161	0.481	0.000	0.000	2.979	0.000	4	-0.079
Y_5	0.369	9.716	0.158	3.000	26.313	1.777	8	-0.344
O_1	3.011	53.349	24.458	4.181	17.721	2.247	22	-2.743
O_2	2.941	16.617	15.304	3.807	5.650	2.435	18	-7.875

Journal of Consciousness Exploration & Research| December 2010 | Vol. 1 | Issue 9 | pp. 1-69
Conte, E., Todarello, O., Conte, S., Mendolicchio, L., Mendolicchio, L. & Federici, A.
Methods and Applications of Non-Linear Analysis in Neurology and Psycho-physiology

46

O_3	1.027	10.087	6.927	3.773	9.824	1.984	16	-1.119
O_5	1.099	11.384	19.940	3.850	10.356	2.524	15	-0.744
O_6	1.389	21.329	33.343	4.089	15.351	2.835	30	-1.452
Vt_3-13	9.354	81.594	88.203	9.482	8.723	4.086	185	-21.035
Vt_2-67	6.137	72.819	78.434	11.743	11.865	4.140	85	-1.163
Vt_1-26	3.294	63.040	75.824	6.100	19.136	3.783	99	-3.408
Vt_1-15	22.430	94.955	96.215	42.409	4.233	6.057	617	-70.089
Vt_1-03	12.528	87.955	91.904	15.395	7.021	4.632	281	-20.9
Vf_2-30	3.756	37.225	55.763	15.284	9.911	3.481	94	-7.896
Vf_2-71	0.371	2.446	0.159	3.000	6.598	1.677	9	0.092
Vf_1-8013	1.173	6.098	8.214	3.852	5.197	1.476	12	0.117
Vf_1-217	20.904	15.356	0.018	4.750	0.735	1.929	21	-0.883
Vf_1-115	18.555	96.544	97.762	40.177	5.203	5.722	557	-59.495

Table 4 Statistical analysis of RQA results (t-Test)						
% DET						
	Yi vs Oi			Oi vs Vfi		
		P value	0.2245		P value	0.6504
		P value summary	ns		P value summary	ns
		Are means signif. different? (P < 0.05)	No		Are means signif. different? (P < 0.05)	No
		t, df	t=1.316 df=8		t, df	t=0.4708 df=8
	Oi vs Vti			Vti vs Vfi		
		P value	0.0004		P value	0.0287
		P value summary	***		P value summary	*
		Are means signif. different? (P < 0.05)	Yes		Are means signif. different? (P < 0.05)	Yes
		t, df	t=5.911 df=8		t, df	t=2.663 df=8
% Lam						
	Yi vs Oi			Oi vs Vfi		
		P value	0.0021		P value	0.55
		P value summary	**		P value summary	ns
		Are means signif. different? (P < 0.05)	Yes		Are means signif. different? (P < 0.05)	No
		t, df	t=4.476 df=8		t, df	t=0.6241 df=8
	Oi vs Vti			Vti vs Vfi		
		P value	P<0.0001		P value	0.0262
		P value summary	***		P value summary	*
		Are means signif. different? (P < 0.05)	Yes		Are means signif. different? (P < 0.05)	Yes
		t, df	t=11.21 df=8		t, df	t=2.722 df=8
T.T.						
	Yi vs Oi			Oi vs Vfi		
		P value	0.0201		P value	0.2161

Journal of Consciousness Exploration & Research| December 2010 | Vol. 1 | Issue 9 | pp. 1-69   47
Conte, E., Todarello, O., Conte, S., Mendolicchio, L., Mendolicchio, L. & Federici, A.
Methods and Applications of Non-Linear Analysis in Neurology and Psycho-physiology

		P value summary	*		P value summary	ns
		Are means signif. different? (P < 0.05)	Yes		Are means signif. different? (P < 0.05)	No
		t, df	t=2.894 df=8		t, df	t=1.343 df=8
	Oi vs Vti			Vti vs Vfi		
		P value	0.0798		P value	0.7166
		P value summary	ns		P value summary	ns
		Are means signif. different? (P < 0.05)	No		Are means signif. different? (P < 0.05)	No
		t, df	t=2.006 df=8		t, df	t=0.3761 df=8
Ratio						
	Yi vs Oi			Oi vs Vfi		
		P value	0.4308		P value	0.0428
		P value summary	ns		P value summary	*
		Are means signif. different? (P < 0.05)	No		Are means signif. different? (P < 0.05)	Yes
		t, df	t=0.8296 df=8		t, df	t=2.406 df=8
	Oi vs Vti			Vti vs Vfi		
		P value	0.647		P value	0.1523
		P value summary	ns		P value summary	ns
		Are means signif. different? (P < 0.05)	No		Are means signif. different? (P < 0.05)	No
		t, df	t=0.4757 df=8		t, df	t=1.582 df=8
Entropy						
	Yi vs Oi			Oi vs Vfi		
		P value	0.0034		P value	0.5926
		P value summary	**		P value summary	ns
		Are means signif. different? (P < 0.05)	Yes		Are means signif. different? (P < 0.05)	No
		t, df	t=4.101 df=8		t, df	t=0.5572 df=8
	Oi vs Vti			Vti vs Vfi		
		P value	0.0011		P value	0.0968
		P value summary	**		P value summary	ns
		Are means signif. different? (P < 0.05)	Yes		Are means signif. different? (P < 0.05)	No
		t, df	t=4.996 df=8		t, df	t=1.881 df=8
Max Line						
	Yi vs Oi			Oi vs Vfi		
		P value	0.0013		P value	0.2956
		P value summary	**		P value summary	ns
		Are means signif. different? (P < 0.05)	Yes		Are means signif. different? (P < 0.05)	No
		t, df	t=4.859 df=8		t, df	t=1.119 df=8
	Oi vs Vti			Vti vs Vfi		
		P value	0.0437		P value	0.4477
		P value summary	*		P value summary	ns
		Are means signif. different? (P < 0.05)	Yes		Are means signif. different? (P < 0.05)	No

Journal of Consciousness Exploration & Research| December 2010 | Vol. 1 | Issue 9 | pp. 1-69   48
Conte, E., Todarello, O., Conte, S., Mendolicchio, L., Mendolicchio, L. & Federici, A.
Methods and Applications of Non-Linear Analysis in Neurology and Psycho-physiology

		t, df	t=2.393 df=8		t, df	t=0.7984 df=8

Table. 5 Calculation of Variability of R-R signals by CZF method.

Subject	(total variability-sec)	VLF (sec^2)	LF (sec^2)	HF (sec^2)	LF/HF	VLF/(LF+HF)
	VT	VLF	LF	HF		
normal						
Y_1	1.398	0.113	0.306	0.619	0.495	0.123
Y_2	1.783	0.207	0.541	1.072	0.505	0.128
Y_3	1.228	0.087	0.263	0.448	0.588	0.122
Y_4	2.057	0.404	1.006	1.743	0.577	0.147
Y_5	1.239	0.103	0.291	0.520	0.559	0.127
O_1	0.756	0.040	0.102	0.189	0.541	0.136
O_2	0.640	0.009	0.031	0.099	0.314	0.072
O_3	0.711	0.029	0.085	0.179	0.473	0.108
O_5	0.577	0.022	0.058	0.121	0.479	0.120
O_6	0.817	0.044	0.127	0.248	0.512	0.116
Ventricular Tachycardia						
Vt_3-13	3.214	0.072	0.445	1.714	0.260	0.033
Vt_2-67	2.453	0.199	0.705	1.839	0.384	0.078
Vt_1-26	2.818	0.385	1.076	2.221	0.484	0.117
Vt_1-15	5.562	3.397	0.276	2.779	0.099	1.112
Vt_1-03	2.141	0.113	0.439	1.223	0.359	0.068
Ventricular Fibrillation						
Vf_2-30	3.106	0.451	1.272	2.608	0.488	0.116
Vf_2-71	2.833	0.361	0.993	2.110	0.471	0.116
Vf_1-8013	4.439	0.905	2.650	5.709	0.464	0.108
Vf_1-217	6.641	2.461	6.424	12.597	0.510	0.129
Vf_1-115	3.708	0.020	0.118	0.976	0.121	0.018

Table 6. Statistical analysis of results obtained by CZF method (t-Test)

VT						
	Yi vs Oi			Oi vs Vfi		
		P value	0.0011		P value	0.001
		P value summary	**		P value summary	**

Journal of Consciousness Exploration & Research| December 2010 | Vol. 1 | Issue 9 | pp. 1-69   49
Conte, E., Todarello, O., Conte, S., Mendolicchio, L., Mendolicchio, L. & Federici, A.
Methods and Applications of Non-Linear Analysis in Neurology and Psycho-physiology

			Are means signif. different? (P < 0.05)	Yes		Are means signif. different? (P < 0.05)	Yes
			t, df	t=4.980 df=8		t, df	t=5.040 df=8
		Oi vs Vti			Vti vs Vfi		
			P value	0.0032		P value	0.3497
			P value summary	**		P value summary	ns
			Are means signif. different? (P < 0.05)	Yes		Are means signif. different? (P < 0.05)	No
			t, df	t=4.162 df=8		t, df	t=0.9933 df=8
VLF							
		Yi vs Oi			Oi vs Vfi		
			P value	0.0321		P value	0.0956
			P value summary	*		P value summary	ns
			Are means signif. different? (P < 0.05)	Yes		Are means signif. different? (P < 0.05)	No
			t, df	t=2.590 df=8		t, df	t=1.889 df=8
		Oi vs Vti			Vti vs Vfi		
			P value	0.2464		P value	0.9936
			P value summary	ns		P value summary	ns
			Are means signif. different? (P < 0.05)	No		Are means signif. different? (P < 0.05)	No
			t, df	t=1.251 df=8		t, df	t=0.008276 df=8
LF							
		Yi vs Oi			Oi vs Vfi		
			P value	0.0219		P value	0.0817
			P value summary	*		P value summary	ns
			Are means signif. different? (P < 0.05)	Yes		Are means signif. different? (P < 0.05)	No
			t, df	t=2.837 df=8		t, df	t=1.991 df=8
		Oi vs Vti			Vti vs Vfi		
			P value	0.007		P value	0.1665
			P value summary	**		P value summary	ns
			Are means signif. different? (P < 0.05)	Yes		Are means signif. different? (P < 0.05)	No
			t, df	t=3.601 df=8		t, df	t=1.522 df=8
HF							
		Yi vs Oi			Oi vs Vfi		
			P value	0.0188		P value	0.0585
			P value summary	*		P value summary	ns
			Are means signif. different? (P < 0.05)	Yes		Are means signif. different? (P < 0.05)	No
			t, df	t=2.936 df=8		t, df	t=2.205 df=8
		Oi vs Vti			Vti vs Vfi		
			P value	0.0001		P value	0.2159
			P value summary	***		P value summary	ns
			Are means signif. different? (P < 0.05)	Yes		Are means signif. different? (P < 0.05)	No
			t, df	t=6.829 df=8		t, df	t=1.344 df=8

Journal of Consciousness Exploration & Research| December 2010 | Vol. 1 | Issue 9 | pp. 1-69
Conte, E., Todarello, O., Conte, S., Mendolicchio, L., Mendolicchio, L. & Federici, A.
Methods and Applications of Non-Linear Analysis in Neurology and Psycho-physiology

50

Table 7. Statistical analysis of results obtained by CZF method (correlation analysis)

	Correlation	Correlation	Correlation	Correlation	Correlation
	VT vs. VLF	VT vs. LF	VT vs. HF	VT vs. LF/HF	VT vs.VLF/(LF+HF)
Y_i	0.952 (*)	0.951 (*)	0.979 (**)	n.s.	n.s.
O_i	n.s.	0.879 (*)	0.923 (*)	n.s.	n.s.
Vt_i	0.953 (*)	n.s.	n.s.	n.s.	0.948 (*)
Vf_i	0.932 (*)	0.933 (*)	0.949 (*)	n.s.	n.s.

Table 8 Values of Hurst exponent

Subjects	H	D=2-H		Statistical analysis (t-Test)	
Y_1	0.070	1.930		Y_i vs O_i	
Y_2	0.125	1.875		P value	0.0142
Y_3	0.281	1.719		P value summary	*
Y_4	0.059	1.941		Are means signif. different? (P < 0.05)	Yes
Y_5	0.163	1.837		t, df	t=3.121 df=8
O_1	0.350	1.650		O_i vs Vt_i	
O_2	0.223	1.777		P value	0.0059
O_3	0.236	1.764		P value summary	**
O_5	0.319	1.681		Are means signif. different? (P < 0.05)	Yes
O_6	0.425	1.575		t, df	t=3.713 df=8
Vt_3-13	0.150	1.850		O_i vs Vf_i	
Vt_2-67	0.036	1.964		P value	0.0007
Vt_1-26	0.046	1.954		P value summary	***
Vt_1-15	0.240	1.760		Are means signif. different? (P < 0.05)	Yes
Vt_1-03	0.098	1.902		t, df	t=5.347 df=8
Vf_2-30	0.082	1.918		Vt_i vs Vf_i	
Vf_2-71	0.050	1.950		P value	0.4414
Vf_1-8013	0.021	1.979		P value summary	ns
Vf_1-217	0.152	1.848		Are means signif. different? (P < 0.05)	No
Vf_1-115	0.089	1.911		t, df	t=0.8099 df=8

Journal of Consciousness Exploration & Research| December 2010 | Vol. 1 | Issue 9 | pp. 1-69
Conte, E., Todarello, O., Conte, S., Mendolicchio, L., Mendolicchio, L. & Federici, A.
Methods and Applications of Non-Linear Analysis in Neurology and Psycho-physiology

51

Table 9 CZF: Analysis of brain waves from spontaneous EEG

delta <4 Hz	4<teta<8 Hz	8<alfa<12 Hz	12<beta<30 Hz	30<gamma<50 Hz	50-125 Hz
315830.18	1546.41	512.81	511.46	124.08	40.96
345604.54	1537.95	485.86	564.77	158.55	65.03
342601.77	1533.87	593.84	992.27	236.30	100.41
231064.75	1184.40	360.35	439.72	135.65	50.88
269108.24	1412.73	477.23	497.33	143.51	69.65
438748.26	2268.66	775.45	781.64	206.83	96.67
1157817.77	5349.84	1487.23	1438.89	395.36	181.99
770427.70	3858.46	1096.07	1095.69	335.10	127.57
296854.43	1635.37	561.97	592.92	177.03	107.97
420348.11	2272.28	769.46	799.45	243.02	135.93
462992.76	2266.45	694.40	773.47	264.10	105.22
855793.00	3727.05	1128.16	1258.53	474.90	209.62
625474.63	2916.07	871.99	882.63	234.88	108.78
430362.95	2232.10	735.97	829.91	261.08	123.09
979082.92	4177.67	1128.36	1435.87	481.32	175.21
882707.17	3041.39	986.53	1046.34	355.57	146.65

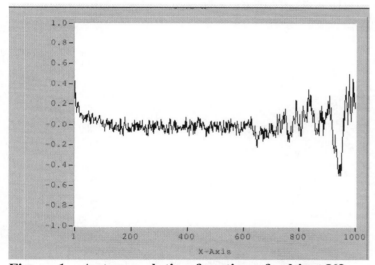

Figure 1a. Autocorrelation function of subject Y2.

Journal of Consciousness Exploration & Research| December 2010 | Vol. 1 | Issue 9 | pp. 1-69
Conte, E., Todarello, O., Conte, S., Mendolicchio, L., Mendolicchio, L. & Federici, A.
Methods and Applications of Non-Linear Analysis in Neurology and Psycho-physiology

52

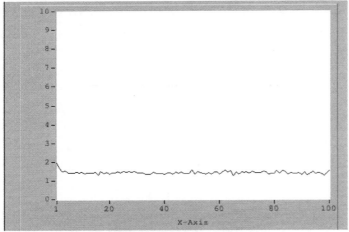

Figure 1b. Mutual Information of subject Y2.

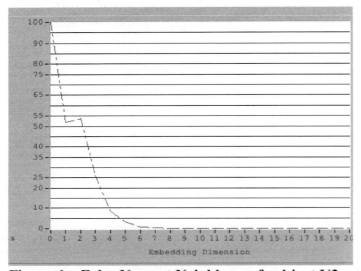

Figure 1c. False Nearest Neighbors of subject Y2.

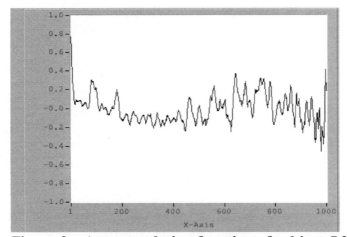

Figure 2a. Autocorrelation function of subject O3.

Journal of Consciousness Exploration & Research| December 2010 | Vol. 1 | Issue 9 | pp. 1-69
Conte, E., Todarello, O., Conte, S., Mendolicchio, L., Mendolicchio, L. & Federici, A.
Methods and Applications of Non-Linear Analysis in Neurology and Psycho-physiology

53

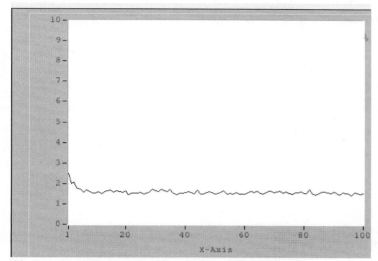

Figure 2b. Mutual Information of subject O3.

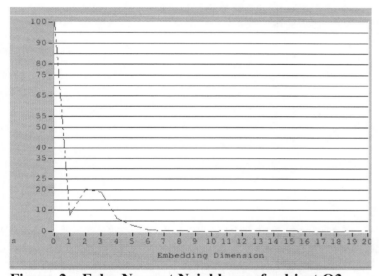

Figure 2c. False Nearest Neighbors of subject O3.

Journal of Consciousness Exploration & Research| December 2010 | Vol. 1 | Issue 9 | pp. 1-69
Conte, E., Todarello, O., Conte, S., Mendolicchio, L., Mendolicchio, L. & Federici, A.
Methods and Applications of Non-Linear Analysis in Neurology and Psycho-physiology

54

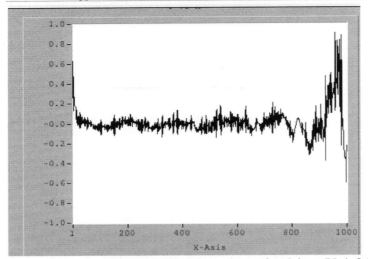

Figure 3a. Autocorrelation function of subject Vt1-26.

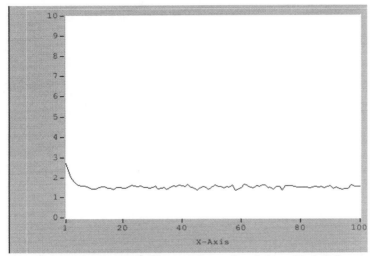

Figure 3b. Mutual Information of subject Vt1-26.

Journal of Consciousness Exploration & Research| December 2010 | Vol. 1 | Issue 9 | pp. 1-69   55
Conte, E., Todarello, O., Conte, S., Mendolicchio, L., Mendolicchio, L. & Federici, A.
Methods and Applications of Non-Linear Analysis in Neurology and Psycho-physiology

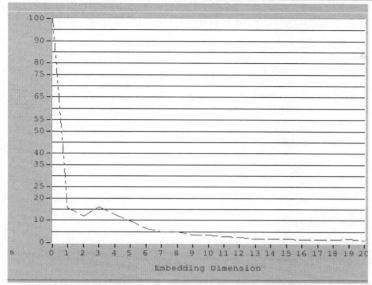

Figure 3c. False Nearest Neighbors of subject Vt1-26.

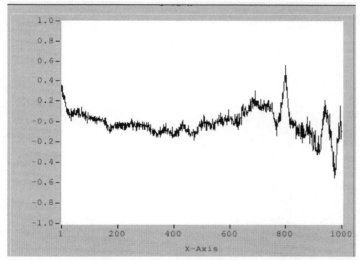

Figure 4a. Autocorrelation function of subject Vf-8013.

Journal of Consciousness Exploration & Research| December 2010 | Vol. 1 | Issue 9 | pp. 1-69
Conte, E., Todarello, O., Conte, S., Mendolicchio, L., Mendolicchio, L. & Federici, A.
Methods and Applications of Non-Linear Analysis in Neurology and Psycho-physiology

56

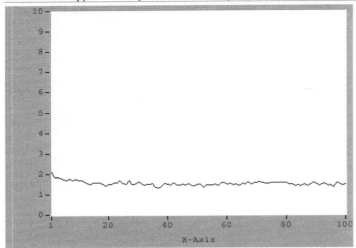

Figure 4b. Mutual Information of subject Vf-8013.

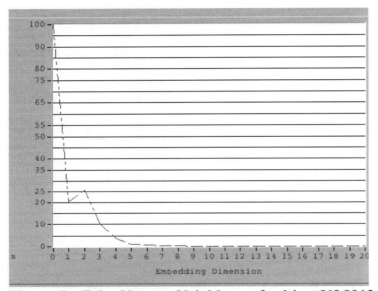

Figure 4c. False Nearest Neighbors of subject Vf-8013.

Journal of Consciousness Exploration & Research| December 2010 | Vol. 1 | Issue 9 | pp. 1-69   57
Conte, E., Todarello, O., Conte, S., Mendolicchio, L., Mendolicchio, L. & Federici, A.
Methods and Applications of Non-Linear Analysis in Neurology and Psycho-physiology

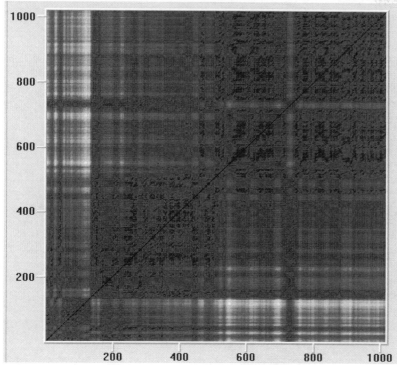

Figure 5. Recurrence Plot of the subject O_2.

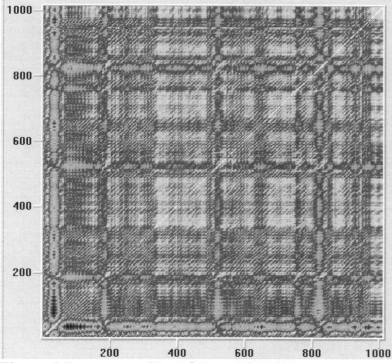

Figure 6. Recurrence Plot of the subject Y_3.

Journal of Consciousness Exploration & Research| December 2010 | Vol. 1 | Issue 9 | pp. 1-69
Conte, E., Todarello, O., Conte, S., Mendolicchio, L., Mendolicchio, L. & Federici, A.
Methods and Applications of Non-Linear Analysis in Neurology and Psycho-physiology

58

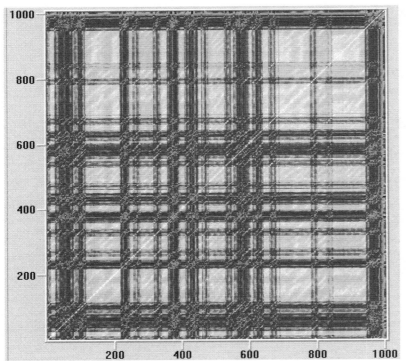

Figure 7. Recurrence Plot of the subject Vt₁-26.

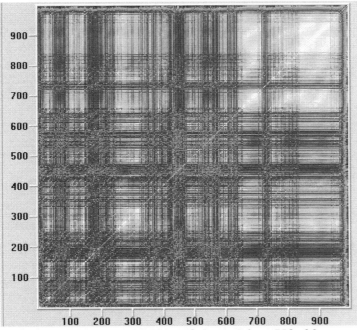

Figure 8. Recurrence Plot of the subject Vf₂-30.

Journal of Consciousness Exploration & Research| December 2010 | Vol. 1 | Issue 9 | pp. 1-69
Conte, E., Todarello, O., Conte, S., Mendolicchio, L., Mendolicchio, L. & Federici, A.
Methods and Applications of Non-Linear Analysis in Neurology and Psycho-physiology

59

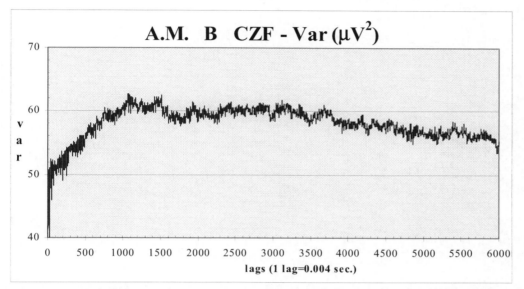

Figure 9. Variability analysis of spontaneous EEG in normal subject (A.M. B)

Figures for analysis of state anxiety

Fig. 1

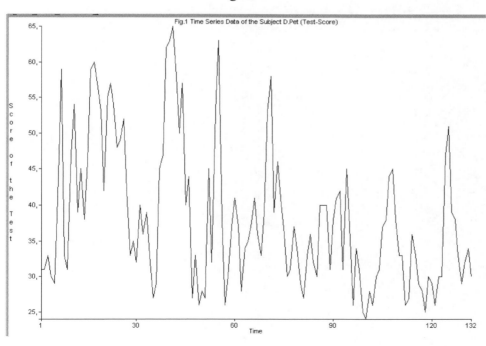

Journal of Consciousness Exploration & Research| December 2010 | Vol. 1 | Issue 9 | pp. 1-69
Conte, E., Todarello, O., Conte, S., Mendolicchio, L., Mendolicchio, L. & Federici, A.
Methods and Applications of Non-Linear Analysis in Neurology and Psycho-physiology

60

Fig.2: Subject F.Dav. - POINCARÉ PLOT
SD1 = 2.11
SD2 = 3.74

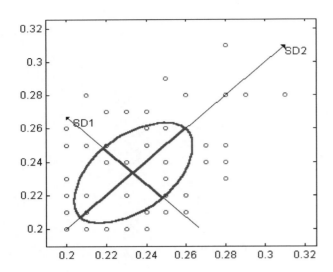

Fig.3: Subject A.Men. - POINCARÉ PLOT
SD1 = 2.21
SD2 = 3.13

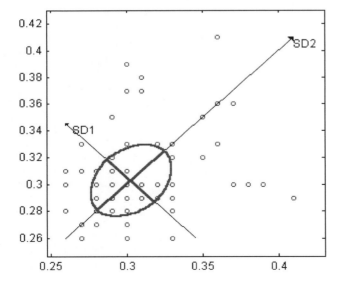

Journal of Consciousness Exploration & Research| December 2010 | Vol. 1 | Issue 9 | pp. 1-69
Conte, E., Todarello, O., Conte, S., Mendolicchio, L., Mendolicchio, L. & Federici, A.
Methods and Applications of Non-Linear Analysis in Neurology and Psycho-physiology

61

Fig.4: Subject A.Mac. - POINCARÉ PLOT
SD1 = 5.46
SD2 = 7.33

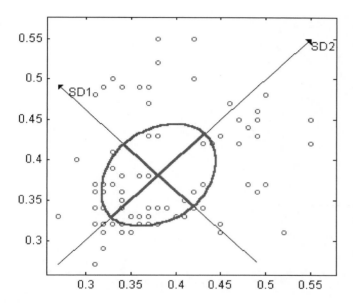

Fig.5: Subject D.Pet. - POINCARÉ PLOT
SD1 = 5.82
SD2 = 12.96

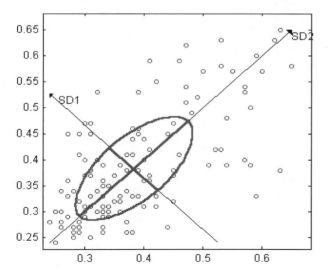

Journal of Consciousness Exploration & Research| December 2010 | Vol. 1 | Issue 9 | pp. 1-69
Conte, E., Todarello, O., Conte, S., Mendolicchio, L., Mendolicchio, L. & Federici, A.
Methods and Applications of Non-Linear Analysis in Neurology and Psycho-physiology

62

Fig.6: Subject M.Den. - POINCARÉ PLOT
SD1 = 9.12
SD2 = 12.64

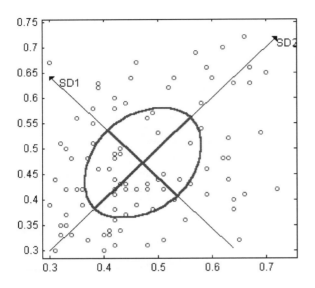

Fig.7: Subject G.Den. - POINCARÉ PLOT
SD1 = 6.89
SD2 = 10.96

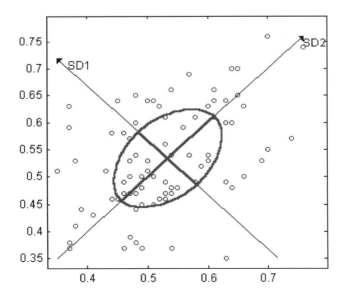

Journal of Consciousness Exploration & Research| December 2010 | Vol. 1 | Issue 9 | pp. 1-69   63
Conte, E., Todarello, O., Conte, S., Mendolicchio, L., Mendolicchio, L. & Federici, A.
Methods and Applications of Non-Linear Analysis in Neurology and Psycho-physiology

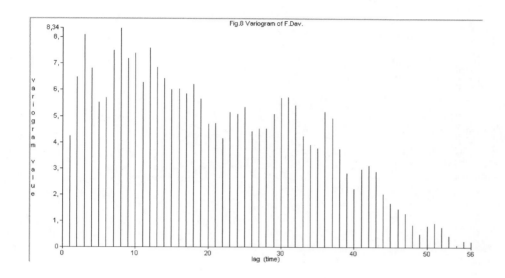

Fig.8 Variogram of F.Dav.

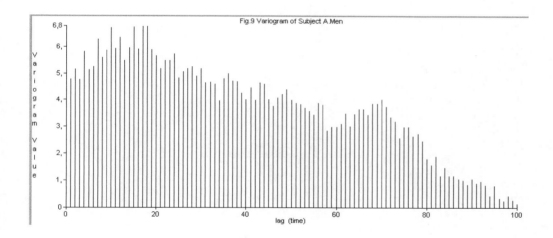

Fig.9 Variogram of Subject A.Men

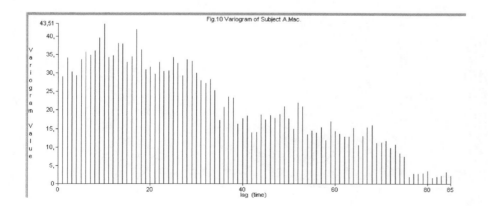

Fig.10 Variogram of Subject A.Mac.

Journal of Consciousness Exploration & Research| December 2010 | Vol. 1 | Issue 9 | pp. 1-69
Conte, E., Todarello, O., Conte, S., Mendolicchio, L., Mendolicchio, L. & Federici, A.
Methods and Applications of Non-Linear Analysis in Neurology and Psycho-physiology

64

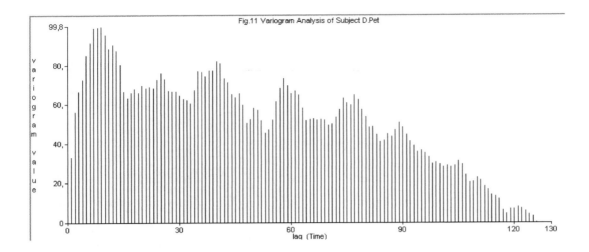

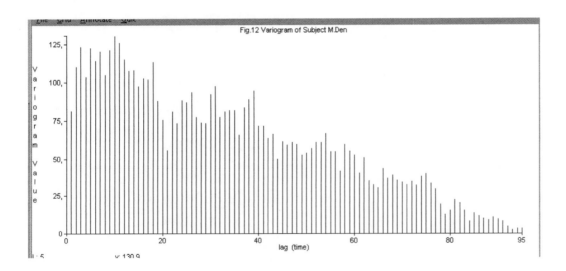

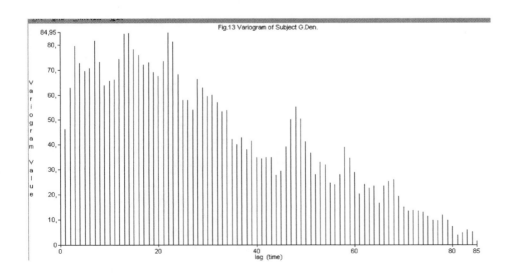

Journal of Consciousness Exploration & Research| December 2010 | Vol. 1 | Issue 9 | pp. 1-69
Conte, E., Todarello, O., Conte, S., Mendolicchio, L., Mendolicchio, L. & Federici, A.
Methods and Applications of Non-Linear Analysis in Neurology and Psycho-physiology

65

Fig. 14 Subject: F. Dav.

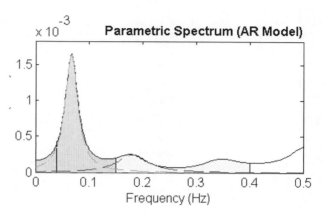

Fig. 15 Subject: A. Men.

Fig. 16 Subject: A. Mac.

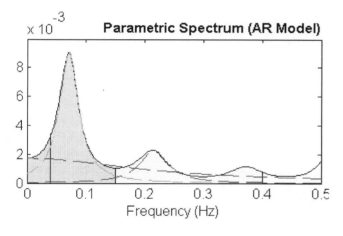

Fig. 17: Subject: D. Pet

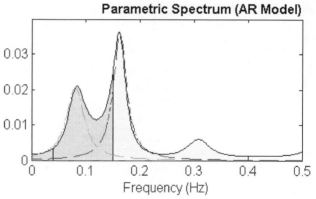

Journal of Consciousness Exploration & Research| December 2010 | Vol. 1 | Issue 9 | pp. 1-69
Conte, E., Todarello, O., Conte, S., Mendolicchio, L., Mendolicchio, L. & Federici, A.
Methods and Applications of Non-Linear Analysis in Neurology and Psycho-physiology

66

Fig.18 Subject M. Den

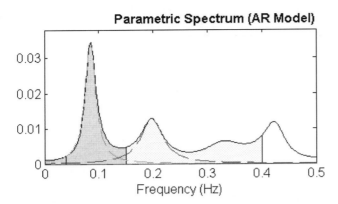

Fig.19: Subject G. Den.

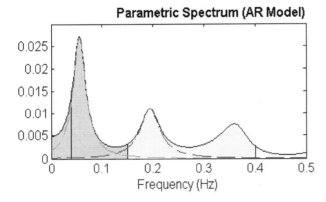

Journal of Consciousness Exploration & Research| December 2010 | Vol. 1 | Issue 9 | pp. 1-69   67
Conte, E., Todarello, O., Conte, S., Mendolicchio, L., Mendolicchio, L. & Federici, A.
Methods and Applications of Non-Linear Analysis in Neurology and Psycho-physiology

Tab. 1 Statistics of Subject F. Dav

Mean	Median
23.62069	23.50000

Maximum	Minimum
31.00000	20.00000

StandardDeviation	Root Mean Squared
2.88174	23.79583

Variance	Skewness
8.30440	0.39860

Kurtosis
-0.82782

Tab. 3 Statistics of Subject G. Den

Mean	Median
53.39773	53.00000

Maximum	Minimum
76.00000	35.00000

StandardDeviation	Root Mean Squared
9.09238	54.16631

Variance	Skewness
82.67136	0.14723

Kurtosis
-0.47710

Tab. 2 Statistics of Subject D. Pet.

Mean	Median
38.31818	36.00000

Maximum	Minimum
65.00000	24.00000

StandardDeviation	Root Mean Squared
9.99494	39.60028

Variance	Skewness
99.89876	0.86033

Kurtosis
-0.12236

Tab. 4 Statistics of Subject A. Mac.

Mean	Median
38.10227	37.00000

Maximum	Minimum
55.00000	27.00000

StandardDeviation	Root Mean Squared
6.39254	38.63480

Variance	Skewness
40.86454	0.79387

Kurtosis
-0.25014

Journal of Consciousness Exploration & Research| December 2010 | Vol. 1 | Issue 9 | pp. 1-69
Conte, E., Todarello, O., Conte, S., Mendolicchio, L., Mendolicchio, L. & Federici, A.
Methods and Applications of Non-Linear Analysis in Neurology and Psycho-physiology

68

Tab. 5 Statistics of Subject A. Men.

Mean	Median
30.28431	30.00000

Maximum	Minimum
41.00000	26.00000

StandardDeviation	Root Mean Squared
2.71300	30.40559

Variance	Skewness
7.36034	1.61347

Kurtosis
2.97502

Tab. 6 Statistics of Subject M. Den.

Mean	Median
47.17526	44.00000

Maximum	Minimum
72.00000	30.00000

StandardDeviation	Root Mean Squared
10.94410	48.42807

Variance	Skewness
119.77341	0.45102

Kurtosis
-0.80755

Tab. 7: SD1 and SD2 Values calculated by Poincaré-Plots

Subject Name	SD1	SD2	Test-Mean Value	Standard Deviation	Variance
F. Dav	2,11	3,74	23,40	2,70	8,30
A. Men	2,21	3,13	30,30	2,40	7,36
A. Mac	5,46	7,33	38,10	6,10	40,86

Journal of Consciousness Exploration & Research| December 2010 | Vol. 1 | Issue 9 | pp. 1-69   69
Conte, E., Todarello, O., Conte, S., Mendolicchio, L., Mendolicchio, L. & Federici, A.
Methods and Applications of Non-Linear Analysis in Neurology and Psycho-physiology

D. Pet	5,82	12,96	38,30	9,00	99,89
M. Den	9,12	12,64	47,20	10,30	119,77
G. Den	6,89	10,96	53,40	8,80	82,67

Tab. 8: Fractal Analysis

Subject Name	Fractal Measure	Generalized Fractal Dimension
F. Dav.	13,200	-0,350
A. Men.	17,500	-0,397
A. Mac.	108,400	-0,450
G. Den.	230,310	-0,495
D. Pet.	269,300	-0,420
M. Den.	305,000	-0,420

Tab. 9   Frequency Domain Analysis

Subject Name	Frequency range (0.1Hz) - Power Spectrum (Test Score)	Frequency range (0.2Hz) - Power Spectrum (Test Score)	Frequency range (0.3-0.4Hz) - Power Spectrum (Test Score)	Frequency range (0.5Hz) - Power Spectrum (Test Score)
F. Dav.	0.00050	0.00300	0.00050	0.00150
A Men.	0.00150	0.00025	0.00018	0.00050
A. Mac.	0.00900	0.00200	0.00180	0.00200
D. Pet.	0.02000	0.03500	0.00500	0.00250
M. Den.	0.03500	0.01500	0.00500	0.01000
G. Den.	0.02500	0.01000	0.00500	0.00010

Journal of Consciousness Exploration & Research| December 2010 | Vol. 1 | Issue 9 | pp. 70-79
Globus, G. *The Dis-closure of World in Waking and Dreaming*

70

Article

The Dis-closure of World in Waking and Dreaming

Gordon Globus[*]

Abstract

The dream world is sometimes indiscernable from the wake world and so parsimony demands a common explanation of world as such. The widespread conviction that the world of dreams is somehow "synthesized" from memory traces of various disparate waking worlds in space and time is implausible, leaving explanation to a mysterious synthetic process. It is argued instead in a Heideggerian vein that closure is fundamental, an "abground" from which world is dis-closed in waking and dreaming both. The abground has dual modes, in accordance with thermofield quantum brain dynamics. Dis-closure of world takes place in the dual modes' belonging-together, whether waking or dreaming. Three factors participate in the process of matching during waking: (1) sensory input representations and their entanglements, (2) self-actions (self-tuning) and their entanglements, (3) re-traces (memory traces of recognitions). The first factor is of course absent during dreaming, yet an authentic world may appear. This leaves waking and dreaming worlds at ontological parity, both dis-closures between-two.

Keywords: Keywords: waking, dreaming, disclosure, memory, ontology, monadology, space & time, abground, thermofield, quantum brain dynamics.

1. INTRODUCTION

There are dream worlds so vivid, so authentically compelling that we have to reason out on waking (sometimes with quaking heart) that it was "only a dream." Indiscernables demand the same explanation on grounds of parsimony, so occasions when the dream world is indistinguishable from the world of waking life raise the questions of how such a feat might be accomplished and its ontological significance.

The great Argentine writer, Jorge Luis Borges, works this theme in a dramatic short story called "The Circular Ruins" (Borges 1998). A man crawls out of the river (out of dynamical flux) into a primal jungle and takes refuge in the circular ruins of an ancient temple destroyed by fire long ago. His main activity becomes dreaming. He is preoccupied in dreaming a son, night by long night, even lovingly dreaming each hair on his son's head and dreaming the son's education

[*] Correspondence: Gordon Globus, Professor Emeritus of Psychiatry and Philosophy, University of California Irvine. Email: ggglobus@uci.edu

lecture by lecture. The man dreams an entire world into which he lovingly places his dream son. The only thing which distinguishes his dream son from a "real" son is that the dream son cannot be harmed by fire. Then one day there is a great conflagration in the jungle and fire sweeps through the circular ruins and engulfs him … only to pass through him without leaving any mark. The man realizes that he, too, is a dream son dreamt by some other!

In this metaphysical drama Borges underlines the indiscernability of waking and dreaming lives. But there is a deficiency in his philosophical argument, which remains metaphysical in spirit. There is always a subjectivity doing the dreaming, a subjectivity who remains superior to world, whilst being dreamt by another subjectivity, who in turn is dreamt by some meta-subject … . Borges' story adopts a traditional Platonic metaphysics or perhaps a Kantian transcendental subjectivity.

It is near-universally believed that memory traces provide a basis for construction of the dream world, which some active compositional or synthesizing process fashions from such raw material. There is disagreement over purpose, mechanism and the relative weight of memory and active process, but the idea that dreaming is at heart constructive on memory materials is not challenged. The title of Hobson and McCarley's (1977) well-known "activation-synthesis" model of dreaming remains relevant today.

The idea of construction from some elements—whether by a subjectivity, a self-organizing dynamics or some other mechanism—is deeply embedded in the mainstream western paradigm. If the dreaming process uses as raw material memory traces from the preceding day (Freud's "day residues") and from the distant past (marked by feeling and unconscious wish according to Freud), then everything is left to some mysterious process of "synthesis" to fashion a seamless dream world from raw materials. Dream synthesis is a vacuous model.

The proposal of Section I will be that the world, whether waking or dreaming, is not a composition or synthesis or construction from something given (sensory input and memory trace) but instead arises in dis-closure. The primal condition is closure (which is a Heideggerian (1999) theme). World, both waking and dreaming, unfolds out of closure. The conditions for world realization are of course different across waking and sleeping but the underlying dynamics remains the same: a belonging-together of dual quantum thermofield theoretical modes. Section II offers detailed analysis of a dream which applies and makes vivid the understanding achieved in Section I. Section III develops the ontological significance of this understanding.

ISSN: 2153-8212   Journal of Consciousness Exploration & Research
                                    Published by QuantumDream, Inc.   www.JCER.com

1. MANIFESTATION OF THE DREAM WORLD

Symmetry and symmetry-breaking

"Symmetry" refers to an indifference to transformation (Rosen 2008; Stewart and Golubitsky 1992). A certain operation makes no difference. Rotate a circle however many degrees, the circle remains unchanged. The circle has "rotational symmetry." Rotate a square 90 degrees and the square is unchanged. Rotations of multiples of 90 degrees also leave the square unchanged; such rotations which do not change the square form a "symmetry group." Rotate the square to any other degree than multiples of 90 and there is change: the symmetry is "broken." Difference appears. Thus a transformation other than 90 degrees or some multiple thereof breaks the rotational symmetry of the square. Symmetry is accordingly a type of plenum—an indifferent plenum of potential differences—which under specific operations may become actual differences.

Water molecules (and many biochemical molecules) within the brain offer a form of symmetry relevant to the present discussion. Water molecules are electric dipoles: one pole is positively charged and the other pole is negatively charged. The difference in charge between poles is represented by a vector arrow, a quantity called the "electric dipole moment vector." When such molecules carry energy, their dipole moment vectors point every which-way. This can be represented by a field with vector arrows at each point, pointing in different directions. Rotate every molecule to the same degree and overall, above the ground state, there is no change: the vectors still point every which-way. Such energized molecules are indifferent to the operation of vector rotation; water molecules at higher energy have dipole rotational symmetry.

The ground state (or vacuum state) of the water dipole field has negligible energy which makes it special. Here the dipoles become correlated, all moment vectors pointing in the same direction. Rotate all the dipoles in the ground state and there is distinct change: the vectors point in a new direction. Rotational symmetry is "broken" in the ground state of the water electric dipole field that spreads through the brain's neural systems.

It is crucial to appreciate that *the region of symmetry-breaking is of macroscopic dimension*. For water molecules it is circa 50 microns (the "coherence length" of water molecules) and these regions may coalesce into larger regions of symmetry-breaking. Different regions communicate via soliton signals in the nanolevel filamentous web of protein fibers that percolate through the brain. (The oft-mentioned water-filled "microtubules" within the neuron and neuroglia are the innermost portion of this nanolevel filamentous web.)

As a consequence of Goldstone's theorem, symmetry lost in the ground state is preserved by a boson condensate of coherent particles. When sensory input dissipates its energy and falls into the ground state, symmetry is broken and a Nambu-Goldstone or "symmetron" boson condensate is formed. These symmetrons provide a *trace* that codes the sensory input. Symmetry-breaking thus supports a quantum physics-based theory of the brain's memory in the form of symmetrons. This theory is called "quantum brain dynamics." For this theory memory traces are ground state condensates of coherent bosons that code sensory input in virtue of preserving the symmetry broken by that input.

A most remarkable feature of this process is that memory as traces of broken symmetry is *total* (Jibu and Yasue 1995). Each new trace is not an added memory element but is *convolved with all preceding traces*. (This total memory is naturally subject to memory loss by spontaneous quantum tunneling processes.) Each new memory is superposed with all memories that came before. The totality of the brain's memory increases with each passing moment. Particular traces enfolded to total memory are weighted by repetition such that total memory becomes sedimented.

So quantum brain dynamics exploits a vacuum state of the symmetry-breaking kind for a theory of sedimented total memory (Jibu and Yasue (1995), Vitiello (1995, 2001). Without symmetry-breaking there would be no memory trace, only imperturbable indifference.

Thermofield quantum brain dynamics

"Dissipative systems" have the ability to decrease their entropy, that is, increase their order, for a time. This is accompanied by a proportional increase in the entropy (decreased order) of the surrounding environment ("heat bath"). In strict principle the "surrounding environment" is not the immediate environs of the dissipative system but the rest of the universe. Thermofield brain dynamics carries this idea to the extreme: here the "system" is *our* universe and the "environment" is an unreachable *alter* universe. The two universes share the ground state; they are "modes" of the ground state. The ground state is "between-two," between two universes, one ours and the other its heat bath which is an alter universe. The arrow of time points forward in our universe and backward in the alter universe. Thermofield dynamics (which adds a true thermodynamical degree of freedom to quantum field theory) thus has dual modalities. The ground state is *between-two*. Dissipative thermofield quantum brain dynamics exploits these dual modes.

The traces of broken symmetry in the ground state of the brain's water electric dipole field are dual mode traces. Each trace is mathematically imaginary, of the form a+bi and a-bi respectively, where $i^2 = -1$. The matching of the dual modes in the ground state is real: $a^2 + b^2$ (since the imaginary terms vanish when complex conjugates are multiplied). A mathematically imaginary dual modes unfolds the mathematically real in belonging-together. To emphasize this movement from imaginary to real, it is called here "dis-closure." Phenomenal reality as world, concrete Being, is dis-closed between-two in the rich ontology supported by thermofield quantum brain dynamics.

The frequent usage of hyphens here is not an affectation. The hyphen in 'between-two' reminds of the tie: Neither mode exists without the other. The hyphen in 'dis-closure' is meant to remind (akin to Heidegger (1999)) that *the fundamental ontological state is closed and that an action is prerequisite to any opening of Being.* Closure should not be confused with emptiness, nor is it the closedness of a Pandora's box with butterfly beings hidden inside. Closure is full, but "what" closure is full of does not exist as such due to the interpenetration.

Three factors control what might belong-together between-two and so be dis-closed. One constraint is of course sensory signals coming in, dissipating their energy, and falling into the ground state. This constraint also encompasses signals arising from within the body, such as drive stimuli and emotional reactions (*pace* the James-Lange theory of emotions). The beauty of focusing on the dream world is that sensory constraints are ineffectual during sleep, so that other constraints might stand out. That an authentic world might be dis-closed in the absence of sensory constraint seems miraculous, when you pause to think about it. (Descartes, sitting in his study, famously observed that he could not rule out the possibility that he was dreaming.)

A second constraint on the between-two is sedimented total memory, which depends on symmetry-breaking. As a participant in the dynamical process of belonging-together, *total memory is prior to Being*, quite the reverse of the way we ordinarily think about it.

A third constraint on the between-two is signals the waking (and REM-sleeping) brain continuously generates on its own, which dissipate their energy and fall into the ground. This constraint is termed here *self-action*. (Self-action is akin to intentionality, but the latter carries so much philosophical baggage that the term will be avoided here.) Were it not for self-action the brain would be stupidly at the mercy of stimulus and memory. *Self-action is situating.* Self-action *means*. Like memory (and unlike the sensory constraint), self-action is richly sedimented. It may be conceived as a self-tuning which develops within social communities, akin to what Searle (1992) means by the "background" or Husserl's (1960) "horizon." As a constraint on the belonging-together of the between-two, self-action can be thought of as "conditions of satisfaction."

Three constraints on ground state belonging-together have been emphasized thus far. What has been left out of the discussion to this point is the *contingent relationship between self-action and other-action*. Symmetrons evoked by cotemporal self-tuning and other-tuning signals become coherently entangled. Later repetition of either one alone calls up the other. The mechanism of "association" is entanglement. As will be seen in the dream illustration below, entanglements play a crucial role in the mechanism of dream formation.

So self-action and other-action (sensory signals), their contingent entanglements, together with total memory, convolve in the ground state between-two, and world is dis-closed in their optimized match, *waking and dreaming alike*. The dissipative thermofield brain is the epitome of

Journal of Consciousness Exploration & Research| December 2010 | Vol. 1 | Issue 9 | pp. 70-79
Globus, G. *The Dis-closure of World in Waking and Dreaming*

75

control on belonging-together in the between-two. Non-living matter and living but brain-less beings do not have such a capability.

2. A DREAM ILLUSTRATION

To put some flesh on the account provided above, it will be illustrated by one of my own dreams. The level of personal detail provided is necessary to appreciate the subtle processes involved in the making of the dream world, which in turn cast a light on the making of the world in waking life. In my view one dream under intense scrutiny can be more illuminating than methodologically rigorous research investigation on many dream subjects. Conviction develops from self-analysis of dreaming, so the present illustration is properly understood as an invitation to self-reflect on one's own dreams.

The Setting for the Dream

On the night of the dream I was sleeping as usual on a floor mattress but in a new place under construction. Feeling a cold coming on, I had thoughtfully placed a box of tissues on a cinder block which happened to be a distant arm's length from the mattress. During the night I awakened with a monumental sustained attack of sneezing. I stuporously reached out my left arm to find the box of tissues where it is usually kept close by my mattress and my hand hit the floor. As I groped around in the adjacent area to the left, the side of my hand kept making contact with the floor—there were three or four tries—and finally remembering where I was, I desperately reached way out to grab the box of tissues.

Two days before the dream I had had quite a surprise. Watching with rapt openness a sonogram of my pregnant wife's abdomen, a rapidly beating heart came into view … and then another rapidly beating heart. Twins! I was barely reconciled to being father of one at my advanced age and two seemed quite a burden. Two days later I still felt unnerved about the startling pair of hearts.

On the day before the dream, in recounting the sonogram experience to some colleagues, they had teased me about my potency in fathering twins (which secretly pleased me).

The Dream

I am operating a bizarre, but within the dream perfectly authentic, piece of medical equipment. There are five or six padded levers sticking out from the front of the machine and two more sticking out from the left side of the machine. [I shall designate these levers from right to left by the numbers one to eight.] I operate the machine by pounding in sequence the padded levers with the downside of my left fist. The sequence of poundings go: 1,2,1,2; 3,4,5, maybe 6; 7,8,7,8. I

Journal of Consciousness Exploration & Research| December 2010 | Vol. 1 | Issue 9 | pp. 70-79
Globus, G. *The Dis-closure of World in Waking and Dreaming*

76

keep repeating this sequence but can't seem to get it right. The first two and last two movements are especially vivid.

Dream Discussion

The dream movements have a family resemblance to my earlier befuddled attempts to find the box of tissues, but is certainly no literal copy of them. The trace of an awkward reaching straight out becomes an awkward reaching way around the corner of the machine. Not the actual reaching with my left arm on mid-sleep awakening but the *eidos* of reaching with it is embodied in the dream, an abstract structure perfectly fitted to the dream circumstances. The reaching on awakening might have infinitely many dream instantiations but in this seamless dream experience there is reaching around the corner of a machine. The urgent self-action involved in reaching for the tissues during waking is vividly repeated during dreaming but since the world is different—the world of operating a piece of medical equipment rather than the world of finding the tissue box—the implementation of reaching is different. Self-tuned reaching participates with other self-actions (as will be seen) in disclosing the dream world.

The theme of the *pair* stands out in the dream, as it did so prominently in my waking life. (The emphatic beginning and end movements each consist in pairs of movements and the pairs are themselves paired.) The day residues have left me self-tuned for pairings; my residual meanings that are poised for activation during REM sleep are dominated by the unexpected repetition of the beating heart, by *two*. The dream expression of this, however, has the repetition superimposed on a movement (1,2,1,2…7,8,7,8). An abstract constraint is imposed on the dreaming process: parse in twos.

A subtle point: The "maybe 6" is not what was actually done in waking (there either was or was not that many gropings) but is based on the uncertain memory of what had actually been done. Thus the authentic dream movements are no concrete revival of waking acts. Instead something *abstract* is at work, the ideas of twoness and hitting with the hand and reaching. There is no conceivable mechanism that might smoothly synthesize memory traces of two hearts, with memory traces of the concrete movement after sneezing, with memory traces of a machine never before seen or even conceived. Freud's "composition" and Hobson and McCarley's "synthesis" are empty processes of explanatory convenience.

What of the bizarre piece of medical equipment, which I had certainly never before seen? There was nothing about it that would identify it as "medical" but in the dream I "somehow knew" (the phrasing is typical for dream recountings) that it was a piece of medical equipment. That is, the self-action *meant* the machine as a piece of medical equipment. The emotionally-charged ("cathected") meaning of "medical equipment" was of course carried over from the sonogram setting and participated in self-tuning under the condition of REM sleep.

The most striking and bizarre feature of the machine were the padded levers which I was *hitting* with the side of my fist. This movement is a gesture of frustration and anger ("pounding the table") at what I initially perceived to be an unwanted burden. I am frustratedly self-tuned. But significantly, on awakening I immediately associated these peculiar levers in the dream with a machine I had in fact seen at the annual traveling carnival when I was a boy. The machine was a test of manly strength and there was a tall analog scale attached for all spectators to see. If the *single* padded lever was pounded by such a powerful fist that the pointer rose to the top of the scale, one was labeled a He-man, or if it barely rose, an unmanly weakling. I had been too fearful of ridicule to ever actually try out my strength on the apparatus. The narcissistic meaning of public potency was embedded in the day residue—father of twins!—provoked by my colleagues' humor. The public teasing about masculine power is entangled with the buried but still cathected boyhood memory of a potency symbol.

So all these meanings (and many more) participate in generating the seamless dream world by means of self-action. The diffuse activation of REM sleep enables emotionally salient day residues (such as medical equipment, twinning, joking colleagues) and their memory entanglements (such as the carnival machine), resulting in dis-closure of an authentic dream life inhabiting an authentic dream world as result. Even the irresolution in the dream experience—I keep repeating the movements and can't seem to get them right—mirrors my inability to regain composure in waking life after the unexpected (but ultimately felicitous) "trauma." Unsettledness is an operator on my dream existence just as it was an operator on my waking existence.

3. ONTOLOGICAL PARITY OF THE WORLD IN WAKING AND DREAMING

Indiscernables demand the same explanation—and the dream world is sometimes indiscernable from the world of waking life for at least some people. It is difficult to focus on this fact and draw proper conclusions from it, so powerful is the impression of world. It seems indubitable that the dream world is ephemeral, totally whisked away in waking up and replaced by the good old solid world of our familiar bedroom. We assume the quotidian world is there whether we are waking, sleeping or dreaming (however much we may "take" it in different ways). "Men come and go. The earth abides." The world is just *so real* and we are absorbed in making our way through world. And our acceptance of world as it seems is vindicated by the outcome: we survive and reproduce. That thermofield dynamics—at least in the version presented here—calls for full parity of waking and dreaming worlds sounds ridiculous. Only mystics, sorcerers, acid heads, and otherwise ignorable *aficionados* of the dream might believe in ontological parity of waking and dreaming worlds!

The present proposal is that world is a formative creation between-two in the ground state, waking and dreaming alike. Even the metrics of space and time are stretched at every moment. (Cf. Heidegger (1982) on temporality and being-in-the-world.) Dis-closure of world in waking is a function of a best match: a belonging-together of (1) self-action (self-tuning) and its entanglements, (2) other-action (sensory input) and its entanglements, and (3) retraces (memory of recognitions). Dis-closure of world in dreaming is also a function of a best match, but a belonging-together of only self-action, its entanglements, and retraces. In dreaming there is no sensory constraint on dis-closure, but world still appears (often a fantastical world to be sure, or a hazy world, but still a world which on occasion is indiscernable from the world of waking life). Self-analysis of dreams convinces that the fresh and authentic dream world could not be seamlessly synthesized from memory snippets.

The term 'consciousness' has been conspicuously absent from the discussion thus far, consistent with the radical ontology proposed. The ontological dichotomy implicit to "consciousness of world" is overcome by primordial closure and through self-action dis-closure. Consciousness is succeeded by *Existenz*, in which we always find ourselves already decisively thrown amidst some world or other in virtue of the dual mode belonging-together.

If world is thus between-two both waking and dreaming, then there is no one world-in-common but instead many worlds in parallel,[1] hoisted at unique addresses wherever there is a rich enough dissipative quantum thermofield system to support self-action, other-action and retrace. Their confluent process is disclosive in belonging-together. To the extent that our inputs, intentions and memory retraces are similar, the worlds amidst which we find ourselves thrown are in agreement. As for the unworldly rest, "it" is closed, distinctionless, a "holomovement" (Bohm 1980) or "abground" (Heidegger 1999). Enchanted by the Goddess *Māyā*, we trust in a world-in-common, erect false ontologies upon it and misconceive our place in the cosmic scheme. The seemingly ephemeral and delusional dream indeed serves as *Via Regia* to the foundations of ontology.

REFERENCES

Bohm, D. (1980) *Wholeness and the implicate order*. London: Routledge & Kegan Paul.
Borges, J.L. (1998) "The circular ruins." In: *Collected Fictions*. A. Hurley, trans. London: Penguin.
Freud (1900) *The interpretation of dreams*. Stanard Edition, v.4-5. J. Strachey, trans.

----

[1] This position should be sharply distinguished from the monadology of Leibniz, for whom there exists a God-given transcendent world operating under efficient causality, in addition to the parallel worlds of monads. The appetitive (attuned) monads each hoist their own version of the transcendent world created by God's "fulgurations."

London: Hogarth Press.

Globus, G. (1987) *Dream life, wake life.* Albany: State University of New York Press.

Globus, G. (2009) *The transparent becoming of world. A crossing between process philosophy and quantum neurophilosophy.* Amsterdam: John Benjamins.

Heidegger, M. (1982 (1927)) *The basic problems of phenomenology.* A. Hofstadter, trans. Bloomington: Indiana University Press.

Heidegger, M. (1999) *Contributions to philosophy: From Enowning.* P. Emad & K. Maly, trans. Bloomington: Indiana University Press.

Hobson, A., McCarley, R.W. (1977) "The brain as a dream-state generator: An activation-synthesis hypothesis." *American J. of Psychiatry* 134: 1335-1368.

Husserl, E. (1960) *Cartesian Meditations.* D. Cairns, trans. The Hague: Nijhoff.

Jibu, M. and Yasue, K. (1995) *An introduction to quantum brain dynamics.* Amsterdam: John Benjamins.

Rosen, J. (2008) *Symmetry rules.* Berlin: Springer Verlag.

Searle, J. R. (1992) *The rediscovery of mind.* Cambridge: MIT Press.

Stewart, I., Golubitsky, M. (1992) *Fearful symmetry.* Oxford: Blackwell.

Vitiello, G. (1995) "Dissipation and memory capacity in the quantum brain model." *International J. of Modern Physics B* 9: 973-989.

Vitiello, G. (2001) *My double unveiled.* Amsterdam: John Benjamins.

ISSN: 2153-8212   Journal of Consciousness Exploration & Research
                                         Published by QuantumDream, Inc.   www.JCER.com

The Principle of Existence II:

Genesis of Self-Referential Matrix Law,

& the Ontology & Mathematics of Ether

Huping Hu[*] & Maoxin Wu
(Dated: December 21, 2010)

ABSTRACT

In the beginning there was Consciousness (prespacetime) by itself $e^0 = 1$ materially empty and spiritually restless. And it began to imagine through primordial self-referential spin $1 = e^{i0} = e^{i0}e^{i0} = e^{iL-iL}e^{iM-iM} = e^{iL}e^{iM}e^{-iL}e^{-iM} = e^{-iL}e^{-iM}/e^{-iL}e^{-iM} = e^{iL}e^{iM}/e^{iL}e^{iM}$...such that it created the self-referential Matrix Law, the external object to be observed and internal object as observed, separated them into external world and internal world, caused them to interact through said Matrix Law and thus gave birth to the Universe which it has since passionately loved, sustained and made to evolve. In short, this work is the continuation of our hypothesis of scientific genesis, sustenance & evolution of the Universe and all creations within (the principle of existence).

Key Words: Conciousness, prespacetime, hierarchical, spin, self-reference, ether, mathematics, ontology, Matrix Law, Transcendental Law of One, Dual-world Law of Zero, Immanent Law of Conservation.

1. INTRODUCTION

Through all of us Consciousness manifests

The beauty and awe of what we continuously discover (or rather what Consciousness is revealing in continuation) is still so ecstatic and the first author is struggling to put them in writing (also see Hu & Wu, 2001-2010). Again, let fellow truth seekers and dear readers be aware that we as humans can only strive for perfection, completeness and correctness in our comprehensions and writings because we ourselves are limited and imperfect.

As shown in our previous work and further revealed in this work, the principles and mathematics which Consciousness may have used to create, sustain and makes evolving of elementary particles and thus the Universe are beautiful and simple.

*Corresponding author: Huping Hu, Ph.D., J.D. Address: QuantumDream, Inc., P.O. Box 267, Stony Brook, NY 11790, USA.
E-mail: hupinghu@quantumbrain.org   Note: the models described herein are the subject of an US patent application (App. No. 12/973,633) filed with USPTO on 12/20/2010.

First, as proposed in the principle of existence, Consciousness employs the following ontological principles among others:

(1) Principle of oneness/unity of existence through quantum entanglement in the body (ether) of prespacetime.

(2) Principle of hierarchical primordial self-referential spin creating:

- Energy-Momentum-Mass Relationship as Transcendental Law of One
- Energy-Momentum-Mass Relationship as Determinant of Matrix Law
- Dual-world Law of Zero of Energy, Momentum & Mass
- Immanent Law of Conservation of Energy, Momentum & Mass in External/Internal World which may be violated in certain processes

Second, as proposed in the principle of existence, Consciousness employs the following mathematical elements & forms among others in order to empower the above ontological principles among others:

(1) e, Euler's number, for (to empower) ether (aether) as foundation/basis/medium of existence (body of prespacetime);
(2) i, imaginary number, for (to empower) thoughts and imagination;
(3) 0, zero, for (to empower) emptiness/undifferentiated/primordial state;
(4) 1, one, for (to empower) oneness/unity of existence;
(5) +, -, *, /, = for (to empower) creation, dynamics, balance & conservation;
(6) Pythagorean theorem for (to empower) Energy-Momentum-Mass Relationship; and
(7) M, matrix, for (to empower) the external and internal worlds (the Dual World) and the interaction of external and internal worlds.

This work is organized as follows. In § 2, we shall illustrate scientific genesis in a nutshell which incorporates the genesis of self-referential Matrix Law. In § 3, we shall detail the genesis of self-referential Matrix Law in the order of: (1) Genesis of Fundamental Energy, Momentum & Mass Relationship; (2) Self-Referential Matrix Law and Its Metamorphoses; (3) Imaginary Momentum; (4) Games for Deriving Matrix Law; and (5) Hierarchical Natural Laws. In § 4, we shall incorporate the genesis of self-referential Matrix Law into scientific genesis of primordial entities (elementary particles) and scientific genesis of composite entities. In § 5, we shall show the mathematics and ontology of ether in the principle of existence. Finally, in § 6, we shall conclude this work. §6 are followed by a dedication and [self-]references.

2. SCIENTIFIC GENESIS IN A NUTSHELL

Consciousness Created Everything
By Self-referential Spin

In the beginning there was Consciousness (prespacetime) by itself $e^0 = 1$ materially empty and spiritually restless. And it began to imagine through primordial self-referential spin

$1=e^{i0}=e^{i0}e^{i0}=e^{iL-iL}e^{iM-iM}=e^{iL}e^{iM}e^{-iL}e^{-iM}=e^{-iL}e^{-iM}/e^{-iL}e^{-iM}=e^{iL}e^{iM}/e^{iL}e^{iM}$ ... such that it created the self-referential Matrix Law, the external object to be observed and internal object as observed, separated them into external world and internal world, caused them to interact through said Matrix Law and thus gave birth to the Universe which it has since passionately loved, sustained and made to evolve.

We draw below several diagrams illustrating the above processes:

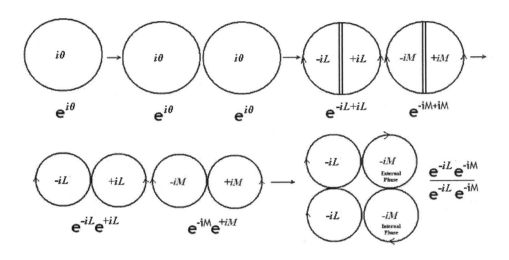

Figure 2.1 Illustration of primordial phase distinction

The primordial phase distinction in Figure 2.1 is accompanied by matrixing of Consciousness body **e** into: (1) external and internal wave functions as external and internal objects, and (2) self-acting and self-referential Matrix Law, which accompany the imaginations of Consciousness head so as to enforce (maintain) the accounting principle of conservation of zero, as illustrated in Figure 2.2.

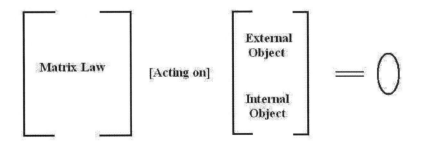

Figure2.2 Consciousness Equation

Figure 2.3 shows from another perspective of the relationship among external object, internal object and the self-acting and self-referential Matrix Law. According to our ontology, self-interactions (self-gravity) are quantum entanglement between the external object and the internal object.

Figure2.3 Self-interaction between external and internal objects of a quantum entity

Therefore, in the principle of existence Consciousness creates, sustains and causes evolution of primordial entities (elementary particles) in prespacetime, that is, within Consciousness itself, by self-referential spin as follows:

$$1 = e^{i0} = e^{i0}e^{i0} = e^{-iL+iL}e^{-iM+iM} = L_e L_i^{-1}\left(e^{-iM}\right)\left(e^{-iM}\right)^{-1} \rightarrow$$

$$\begin{pmatrix} L_{M,e} & L_{M,i} \end{pmatrix}\begin{pmatrix} A_e e^{-iM} \\ A_i e^{-iM} \end{pmatrix} = L_M \begin{pmatrix} A_e \\ A_i \end{pmatrix}e^{-iM} = L_M \begin{pmatrix} \Psi_e \\ \Psi_i \end{pmatrix} = L_M \psi = 0 \qquad (2.1)$$

In expression (2.1), e is Euler number representing Consciousness body (ether or aether), i is imaginary unit representing Consciousness' imagination in Consciousness head, $\pm M$ is immanent content of imagination i such as space, time, momentum & energy, $\pm L$ is immanent law of imagination i, $L_1 = e^{i0} = e^{-iL+iL} = L_e L_i^{-1} = 1$ is Consciousness'

Transcendental Law of One before matrixization, L_e is external law, L_i is internal law, $L_{M,e}$ is external matrix law, and $L_{M,i}$ is internal matrix law, L_M is Consciousness' self-referential Matrix Law comprised of external and internal matrix laws which governs elementary entities and conserves zero, Ψ_e is external wave function (external object), Ψ_i is internal wave function (internal object)and Ψ is the complete wave function (object/entity in the dual-world as a whole).

Consciousness spins as $1 = e^{i0} = e^{i0}e^{i0} = e^{iL-iL}e^{iM-iM} = e^{iL}e^{iM}e^{-iL}e^{-iM} = e^{-iL}e^{-iM}/e^{-iL}e^{-iM} = e^{iL}e^{iM}/e^{iL}e^{iM}$
…before matrixization. Consciousness also spins through self-acting and self-referential Matrix Law L_M after matrixization which acts on external object and internal object to cause them to interact with each other as further described below.

3. GENESIS OF SELF-REFERENTIAL MATRIX LAW

Natural laws are hierarchical

3.1 Genesis of Fundamental Energy, Momentum & Mass Relationship

In the principle of existence, fundamental energy, momentum & mass relationship:

$$E^2 = m^2 + \mathbf{p}^2 \quad \text{or} \quad E^2 - m^2 - \mathbf{p}^2 = 0 \tag{3.1}$$

is created from the following primordial self-referential spin:

$$1 = e^{i0} = e^{-iL+iL} = L_e L_i^{-1} = (\cos L - i \sin L)(\cos L + i \sin L) =$$

$$\left(\frac{m}{E} - i \frac{|\mathbf{p}|}{E} \right)\left(\frac{m}{E} + i \frac{|\mathbf{p}|}{E} \right) = \left(\frac{m - i|\mathbf{p}|}{E} \right)\left(\frac{m + i|\mathbf{p}|}{E} \right) = \left(\frac{m^2 + \mathbf{p}^2}{E^2} \right) \rightarrow$$

$$E^2 = m^2 + \mathbf{p}^2 \tag{3.2}$$

For simplicity, we have set $c=1$ in equation (3.4) and will set $c=\hbar=1$ through out this work unless indicated otherwise. Expression (3.4) was discovered by Einstein.

In the presence of an interacting field of a second primordial entity such as an electromagnetic potential:

$$A^\mu = (\phi, \mathbf{A}) \tag{3.3}$$

equation (3.4) becomes for an elementary entity with electric charge e:

$$1 = e^{i0} = e^{-iL+iL} = L_e L_i^{-1} = (\cos L - i \sin L)(\cos L + i \sin L) =$$

$$\left(\frac{m}{E - e\phi} - i \frac{|\mathbf{p} - e\mathbf{A}|}{E - e\phi} \right)\left(\frac{m}{E - e\phi} + i \frac{|\mathbf{p} - e\mathbf{A}|}{E - e\phi} \right) =$$

$$\left(\frac{m - i|\mathbf{p} - e\mathbf{A}|}{E - e\phi} \right)\left(\frac{m + i|\mathbf{p} - e\mathbf{A}|}{E - e\phi} \right) = \left(\frac{m^2 + |\mathbf{p} - e\mathbf{A}|^2}{(E - e\phi)^2} \right) \rightarrow$$

$$(E - e\phi)^2 = m^2 + (\mathbf{p} - e\mathbf{A})^2 \quad \text{or} \quad (E - e\phi)^2 - m^2 - (\mathbf{p} - e\mathbf{A})^2 = 0 \tag{3.4}$$

3.2 Self-Referential Matrix Law and Its Metamorphoses

In the principle of existence, one form of Consciousness' Matrix Law L_M is created from the following primordial self-referential spin:

$$1 = e^{i0} = e^{-iL+iL} = L_e L_i^{-1} = (\cos L - i \sin L)(\cos L + i \sin L) =$$

$$\left(\frac{m}{E} - i \frac{|\mathbf{p}|}{E} \right)\left(\frac{m}{E} + i \frac{|\mathbf{p}|}{E} \right) = \left(\frac{m - i|\mathbf{p}|}{E} \right)\left(\frac{m + i|\mathbf{p}|}{E} \right) = \left(\frac{m^2 + \mathbf{p}^2}{E^2} \right)$$

$$= \frac{E^2 - m^2}{\mathbf{p}^2} = \left(\frac{E - m}{-|\mathbf{p}|} \right)\left(\frac{-|\mathbf{p}|}{E + m} \right)^{-1}$$

$$\rightarrow \frac{E - m}{-|\mathbf{p}|} = \frac{-|\mathbf{p}|}{E + m} \rightarrow \frac{E - m}{-|\mathbf{p}|} - \frac{-|\mathbf{p}|}{E + m} = 0 \qquad (3.5)$$

$$\rightarrow \begin{pmatrix} E - m & -|\mathbf{p}| \\ -|\mathbf{p}| & E + m \end{pmatrix} = \begin{pmatrix} L_{M,e} & L_{M,i} \end{pmatrix} = L_M$$

where matrixization step is carried out in such way that

$$\mathrm{Det}(L_M) = E^2 - m^2 - \mathbf{p}^2 = 0 \qquad (3.6)$$

so as to satisfy the fundamental relationship (3.4) in the determinant view.

After fermionic spinization:

$$|\mathbf{p}| = \sqrt{\mathbf{p}^2} = \sqrt{-Det(\boldsymbol{\sigma} \cdot \mathbf{p})} \rightarrow \boldsymbol{\sigma} \cdot \mathbf{p} \qquad (3.7)$$

where $\boldsymbol{\sigma} = (\sigma_1, \sigma_2, \sigma_3)$ are Pauli matrices:

$$\sigma_1 = \begin{pmatrix} 0 & 1 \\ 1 & 0 \end{pmatrix} \qquad \sigma_2 = \begin{pmatrix} 0 & -i \\ i & 0 \end{pmatrix} \qquad \sigma_3 = \begin{pmatrix} 1 & 0 \\ 0 & -1 \end{pmatrix} \qquad (3.8)$$

expression (3.7) becomes:

$$\begin{pmatrix} E - m & -\boldsymbol{\sigma} \cdot \mathbf{p} \\ -\boldsymbol{\sigma} \cdot \mathbf{p} & E + m \end{pmatrix} = \begin{pmatrix} L_{M,e} & L_{M,i} \end{pmatrix} = L_M = E - \boldsymbol{\alpha} \bullet \mathbf{p} - \beta m = E - H \qquad (3.9)$$

where $\boldsymbol{\alpha} = (\alpha_1, \alpha_2, \alpha_3)$ and β are Dirac matrices and $H = \boldsymbol{\alpha} \bullet \mathbf{p} + \beta m$ is the Dirac Hamiltonian. Expression (3.12) governs fermions in Dirac form such as Dirac electron and positron and we propose that expression (3.7) governs the third state of matter (unspinized or spinless entity/particle) with electric charge e and mass m such as a meson or a meson-like particle.

If we define:

$$Det_\sigma \begin{pmatrix} E-m & -\boldsymbol{\sigma}\cdot\mathbf{p} \\ -\boldsymbol{\sigma}\cdot\mathbf{p} & E+m \end{pmatrix} = (E-m)(E+m) - (-\boldsymbol{\sigma}\cdot\mathbf{p})(-\boldsymbol{\sigma}\cdot\mathbf{p}) \tag{3.10}$$

We get:

$$Det_\sigma \begin{pmatrix} E-m & -\boldsymbol{\sigma}\cdot\mathbf{p} \\ -\boldsymbol{\sigma}\cdot\mathbf{p} & E+m \end{pmatrix} = (E^2 - m^2 - \mathbf{p}^2)I_2 = 0 \tag{3.11}$$

Thus, fundamental relationship (3.1) is also satisfied under the determinant view of expression (3.13). Indeed, we can also obtain the following conventional determinant:

$$Det \begin{pmatrix} E-m & -\boldsymbol{\sigma}\cdot\mathbf{p} \\ -\boldsymbol{\sigma}\cdot\mathbf{p} & E+m \end{pmatrix} = (E^2 - m^2 - \mathbf{p}^2)^2 = 0 \tag{3.12}$$

One kind of metamorphosis of expressions (3.5), (3.9), (310) & (3.11) is respectively as follows:

$$1 = e^{i0} = e^{-iL+iL} = L_e L_i^{-1} = (\cos L - i \sin L)(\cos L + i \sin L) =$$

$$\left(\frac{m}{E} - i\frac{|\mathbf{p}|}{E}\right)\left(\frac{m}{E} + i\frac{|\mathbf{p}|}{E}\right) = \left(\frac{m - i|\mathbf{p}|}{E}\right)\left(\frac{m + i|\mathbf{p}|}{E}\right) = \left(\frac{m^2 + \mathbf{p}^2}{E^2}\right) =$$

$$\frac{E^2 - \mathbf{p}^2}{m^2} = \left(\frac{E - |\mathbf{p}|}{-m}\right)\left(\frac{-m}{E + |\mathbf{p}|}\right)^{-1} \rightarrow$$

$$\rightarrow \frac{E - |\mathbf{p}|}{-m} = \frac{-m}{E + |\mathbf{p}|} \rightarrow \frac{E - |\mathbf{p}|}{-m} - \frac{-m}{E + |\mathbf{p}|} = 0 \tag{3.13}$$

$$\begin{pmatrix} E - |\mathbf{p}| & -m \\ -m & E + |\mathbf{p}| \end{pmatrix} = \begin{pmatrix} L_{M,e} & L_{M,i} \end{pmatrix} = L_M$$

$$\begin{pmatrix} E-\boldsymbol{\sigma}\cdot\mathbf{p} & -m \\ -m & E+\boldsymbol{\sigma}\cdot\mathbf{p} \end{pmatrix} = \begin{pmatrix} L_{M,e} & L_{M,i} \end{pmatrix} = L_M \tag{3.14}$$

$$Det_\sigma \begin{pmatrix} E-\boldsymbol{\sigma}\cdot\mathbf{p} & -m \\ -m & E+\boldsymbol{\sigma}\cdot\mathbf{p} \end{pmatrix} = (E-\boldsymbol{\sigma}\cdot\mathbf{p})(E+\boldsymbol{\sigma}\cdot\mathbf{p})-(-m)(-m) \tag{3.15}$$

$$Det_\sigma \begin{pmatrix} E-\boldsymbol{\sigma}\cdot\mathbf{p} & -m \\ -m & E+\boldsymbol{\sigma}\cdot\mathbf{p} \end{pmatrix} = \left(E^2 -\mathbf{p}^2 - m^2\right)I_2 = 0 \tag{3.16}$$

The last expression in (3.13) is the unspinized Matrix Law in Weyl (chiral) form. Expression (3.14) is spinized Matrix Law in Weyl (chiral) form.

Another kind of metamorphosis of expressions (3.5), (3.9), (310) & (3.11) is respectively as follows:

$$1 = e^{i0} = e^{-iL+iL} = L_e L_i^{-1} = (\cos L - i\sin L)(\cos L + i\sin L) =$$

$$\left(\frac{m}{E} - i\frac{|\mathbf{p}|}{E}\right)\left(\frac{m}{E} + i\frac{|\mathbf{p}|}{E}\right) = \left(\frac{m-i|\mathbf{p}|}{E}\right)\left(\frac{m+i|\mathbf{p}|}{E}\right) = \left(\frac{E}{-m+i|\mathbf{p}|}\right)^{-1}\left(\frac{-m-i|\mathbf{p}|}{E}\right) \tag{3.17}$$

$$\rightarrow \frac{E}{-m+i|\mathbf{p}|} = \frac{-m-i|\mathbf{p}|}{E} \rightarrow \frac{E}{-m+i|\mathbf{p}|} - \frac{-m-i|\mathbf{p}|}{E} = 0$$

$$\rightarrow \begin{pmatrix} E & -m-i|\mathbf{p}| \\ -m+i|\mathbf{p}| & E \end{pmatrix} = \begin{pmatrix} L_e & L_i \end{pmatrix} = L_M$$

$$\begin{pmatrix} E & -m-i\boldsymbol{\sigma}\cdot\mathbf{p} \\ -m+i\boldsymbol{\sigma}\cdot\mathbf{p} & E \end{pmatrix} = \begin{pmatrix} L_{M,e} & L_{M,i} \end{pmatrix} = L_M \tag{3.18}$$

$$Det_\sigma \begin{pmatrix} E & -m-i\boldsymbol{\sigma}\cdot\mathbf{p} \\ -m+i\boldsymbol{\sigma}\cdot\mathbf{p} & E \end{pmatrix} = EE - (-m-i\boldsymbol{\sigma}\cdot\mathbf{p})(-m+i\boldsymbol{\sigma}\cdot\mathbf{p}) \tag{3.19}$$

$$Det_\sigma \begin{pmatrix} E & -m-i\boldsymbol{\sigma}\cdot\mathbf{p} \\ -m+i\boldsymbol{\sigma}\cdot\mathbf{p} & E \end{pmatrix} = \left(E^2 - m^2 - \mathbf{p}^2\right)I_2 = 0 \tag{3.20}$$

Indeed, $Q = m + i\boldsymbol{\sigma}\cdot\mathbf{p}$ is a quaternion and $Q^* = m - i\boldsymbol{\sigma}\cdot\mathbf{p}$ is its conjugate. So we can rewrite expression (3.29) as:

$$\begin{pmatrix} E & -Q \\ -Q^* & E \end{pmatrix} = \begin{pmatrix} L_{M,e} & L_{M,i} \end{pmatrix} = L_M \tag{3.21}$$

If $m=0$, we have from expression (3.5):

$$1 = e^{i0} = e^{-iL+iL} = L_e L_i^{-1} = (\cos L - i \sin L)(\cos L + i \sin L) =$$

$$\left(\frac{0}{E} - i\frac{|\mathbf{p}|}{E}\right)\left(\frac{0}{E} + i\frac{|\mathbf{p}|}{E}\right) = \left(-i\frac{|\mathbf{p}|}{E}\right)\left(+i\frac{|\mathbf{p}|}{E}\right) = \left(\frac{\mathbf{p}^2}{E^2}\right)$$

$$= \frac{E^2}{\mathbf{p}^2} = \left(\frac{E}{-|\mathbf{p}|}\right)\left(\frac{-|\mathbf{p}|}{E}\right)^{-1}$$

(3.22)

$$\rightarrow \frac{E}{-|\mathbf{p}|} = \frac{-|\mathbf{p}|}{E} \rightarrow \frac{E}{-|\mathbf{p}|} - \frac{-|\mathbf{p}|}{E} = 0$$

$$\rightarrow \begin{pmatrix} E & -|\mathbf{p}| \\ -|\mathbf{p}| & E \end{pmatrix} = \begin{pmatrix} L_{M,e} & L_{M,i} \end{pmatrix} = L_M$$

After fermionic spinization $|\mathbf{p}| \rightarrow \boldsymbol{\sigma} \cdot \mathbf{p}$, the last expression in (3.22) becomes:

$$\begin{pmatrix} E & -\boldsymbol{\sigma}\cdot\mathbf{p} \\ -\boldsymbol{\sigma}\cdot\mathbf{p} & E \end{pmatrix} = \begin{pmatrix} L_{M,e} & L_{M,i} \end{pmatrix} = L_M$$

(3.23)

which governs massless fermion (neutrino) in Dirac form.

After bosonic spinization:

$$|\mathbf{p}| = \sqrt{\mathbf{p}^2} = \sqrt{-(Det(\mathbf{s}\cdot\mathbf{p}+I_3)-Det(I_3))} \rightarrow \mathbf{s} \cdot \mathbf{p}$$

(3.24)

the expression in (3.22) becomes:

$$\begin{pmatrix} E & -\mathbf{s}\cdot\mathbf{p} \\ -\mathbf{s}\cdot\mathbf{p} & E \end{pmatrix} = \begin{pmatrix} L_{M,e} & L_{M,i} \end{pmatrix} = L_M$$

(3.25)

where $\mathbf{s} = (s_1, s_2, s_3)$ are spin operators for spin 1 particle:

$$s_1 = \begin{pmatrix} 0 & 0 & 0 \\ 0 & 0 & -i \\ 0 & i & 0 \end{pmatrix} \quad s_2 = \begin{pmatrix} 0 & 0 & i \\ 0 & 0 & 0 \\ -i & 0 & 0 \end{pmatrix} \quad s_3 = \begin{pmatrix} 0 & -i & 0 \\ i & 0 & 0 \\ 0 & 0 & 0 \end{pmatrix}$$

(3.26)

If we define:

$$Det_s \begin{pmatrix} E & -\mathbf{s}\cdot\mathbf{p} \\ -\mathbf{s}\cdot\mathbf{p} & E \end{pmatrix} = (E)(E) - (-\mathbf{s}\cdot\mathbf{p})(-\mathbf{s}\cdot\mathbf{p}) \tag{3.27}$$

We get:

$$Det_s \begin{pmatrix} E & -\mathbf{s}\cdot\mathbf{p} \\ -\mathbf{s}\cdot\mathbf{p} & E \end{pmatrix} = \left(E^2 - \mathbf{p}^2\right)I_3 - \begin{pmatrix} p_x^2 & p_x p_y & p_x p_z \\ p_y p_x & p_y^2 & p_y p_z \\ p_z p_x & p_z p_y & p_z^2 \end{pmatrix} \tag{3.28}$$

To obey fundamental relationship (3.1) in determinant view (3.27), we shall require the last term in (3.28) acting on the external and internal wave functions respectively to produce null result (zero) in source-free zone as discussed later. We propose that the last expression in (3.22) governs massless particle with unobservable spin (spinless). After bosonic spinization, the spinless and massless particle gains its spin 1.

Further, if $|\mathbf{p}|=0$, we have:

$$1 = e^{i0} = e^{-iL+iL} = L_e L_i^{-1} = (\cos L - i\sin L)(\cos L + i\sin L) =$$

$$\left(\frac{m}{E} - i\frac{0}{E}\right)\left(\frac{m}{E} + i\frac{0}{E}\right) = \left(\frac{m}{E}\right)\left(\frac{m}{E}\right) = \left(\frac{m^2}{E^2}\right)$$

$$= \frac{E^2}{m^2} = \left(\frac{E}{-m}\right)\left(\frac{-m}{E}\right)^{-1}$$

$$\rightarrow \frac{E}{-m} = \frac{-m}{E} \rightarrow \frac{E}{-m} - \frac{-m}{E} = 0$$

$$\rightarrow \begin{pmatrix} E & -m \\ -m & E \end{pmatrix} = \left(L_{M,e} \quad L_{M,i}\right) = L_M \tag{3.29}$$

We suggest the above spaceless forms of Matrix Law govern the external and internal wave functions (self-fields) which play the roles of spaceless gravitons, that is, they mediate space (distance) independent interactions through proper time (mass) entanglement.

3.3 Imaginary Momentum

If Consciousness creates spatial self-confinement of an elementary entity through imaginary momentum $\mathbf{p}_i$ (downward self-reference such that $m^2 > E^2$) we have:

$$m^2 - E^2 = -\mathbf{p}_i^2 = -p_{i,1}^2 - p_{i,2}^2 - p_{i,3}^2 = (i\mathbf{p}_i)^2 = -Det(\boldsymbol{\sigma} \cdot i\mathbf{p}_i) \tag{3.30}$$

that is:

$$E^2 - m^2 - \mathbf{p}_i^2 = 0 \tag{3.31}$$

which can be created by the following primordial self-referential spin:

$$1 = e^{i0} = e^{-iL+iL} = L_e L_i^{-1} = (\cos L - i\sin L)(\cos L + i\sin L) =$$

$$\left(\frac{m}{E} - i\frac{|\mathbf{p}_i|}{E}\right)\left(\frac{m}{E} + i\frac{|\mathbf{p}_i|}{E}\right) = \left(\frac{m - i|\mathbf{p}_i|}{E}\right)\left(\frac{m + i|\mathbf{p}_i|}{E}\right) = \left(\frac{m^2 + \mathbf{p}_i^2}{E^2}\right) \rightarrow$$

$$E^2 = m^2 + \mathbf{p}_i^2 \quad \text{or} \quad E^2 - m^2 - \mathbf{p}_i^2 = 0 \tag{3.32}$$

Therefore, allowing imaginary momentum (downward self-reference) for an elementary entity, we can derive the following Matrix Law in Dirac-like form:

$$\begin{pmatrix} E-m & -|\mathbf{p}_i| \\ -|\mathbf{p}_i| & E+m \end{pmatrix} = \begin{pmatrix} L_{M,e} & L_{M,i} \end{pmatrix} = L_M \tag{3.33}$$

$$\begin{pmatrix} -m & -\boldsymbol{\sigma}\cdot\mathbf{p}_i \\ -\boldsymbol{\sigma}\cdot\mathbf{p}_i & +m \end{pmatrix} = \begin{pmatrix} L_{M,e} & L_{M,i} \end{pmatrix} = L_M \tag{3.34}$$

Also, we can derive the following Matrix Law in Weyl-like (chiral-like) form:

$$\begin{pmatrix} E-|\mathbf{p}_i| & -m \\ -m & +|\mathbf{p}_i| \end{pmatrix} = \begin{pmatrix} L_{M,e} & L_{M,i} \end{pmatrix} = L_M \tag{3.35}$$

$$\begin{pmatrix} E-\boldsymbol{\sigma}\cdot\mathbf{p}_i & -m \\ -m & E+\boldsymbol{\sigma}\cdot\mathbf{p}_i \end{pmatrix} = \begin{pmatrix} L_{M,e} & L_{M,i} \end{pmatrix} = L_M \tag{3.36}$$

It is suggested that the above additional forms of self-referential Matrix Law govern proton in Dirac and Weyl form respectively.

3.4 Games for Deriving Matrix Law

The games for deriving various forms of the Matrix Law prior to spinization can be summarized as follows:

$$0 = E^2 - m^2 - \mathbf{p}^2 = \left(DetM_E + DetM_m + DetM_p \right) \qquad (3.37)$$

$$= Det(M_E + M_m + M_p) = Det(L_M)$$

where *Det* means determinant and M_E, M_m and M_p are respectively matrices with $\pm E$ (or $\pm iE$), $\pm m$ (or $\pm im$) and $\pm|\mathbf{p}|$ (or $\pm i|\mathbf{p}|$) as elements respectively, and E^2, $-m^2$ and $-\mathbf{p}^2$ as determinant respectively, and L_M is the Matrix Law so derived.

For example, the Matrix Law in Dirac form prior to spinization:

$$L_M = \begin{pmatrix} E-m & -|\mathbf{p}| \\ -|\mathbf{p}| & E+m \end{pmatrix} \qquad (3.38)$$

can be derived as follows:

$$0 = E^2 - m^2 - \mathbf{p}^2 = Det\begin{pmatrix} E & 0 \\ 0 & E \end{pmatrix} + Det\begin{pmatrix} -m & 0 \\ 0 & m \end{pmatrix} + Det\begin{pmatrix} 0 & -|\mathbf{p}| \\ -|\mathbf{p}| & 0 \end{pmatrix} =$$

$$Det\left(\begin{pmatrix} E & 0 \\ 0 & E \end{pmatrix} + \begin{pmatrix} -m & 0 \\ 0 & m \end{pmatrix} + \begin{pmatrix} 0 & -|\mathbf{p}| \\ -|\mathbf{p}| & 0 \end{pmatrix} \right) = Det\begin{pmatrix} E-m & -|\mathbf{p}| \\ -|\mathbf{p}| & E+m \end{pmatrix} = Det(L_M) \qquad (3.39)$$

For a second example, the Matrix Law in Weyl form prior to spinization:

$$L_M = \begin{pmatrix} E-|\mathbf{p}| & -m \\ -m & E+|\mathbf{p}| \end{pmatrix} \qquad (3.40)$$

can be derived as follows:

$$0 = E^2 - m^2 - \mathbf{p}^2 = Det\begin{pmatrix} E & 0 \\ 0 & E \end{pmatrix} + Det\begin{pmatrix} 0 & -m \\ -m & 0 \end{pmatrix} + Det\begin{pmatrix} -|\mathbf{p}| & 0 \\ 0 & |\mathbf{p}| \end{pmatrix} =$$

$$Det\left(\begin{pmatrix} E & 0 \\ 0 & E \end{pmatrix} + \begin{pmatrix} 0 & -m \\ -m & 0 \end{pmatrix} + \begin{pmatrix} -|\mathbf{p}| & 0 \\ 0 & |\mathbf{p}| \end{pmatrix} \right) = Det\begin{pmatrix} E-|\mathbf{p}| & -m \\ -m & E+|\mathbf{p}| \end{pmatrix} = Det(L_M) \qquad (3.41)$$

For a third example, the Matrix Law in Quaternion form prior to spinization:

$$L_M = \begin{pmatrix} E & -m-i|\mathbf{p}| \\ -m+i|\mathbf{p}| & E \end{pmatrix} \qquad (3.42)$$

can be derived as follows:

$$0 = E^2 - m^2 - \mathbf{p}^2 = Det\begin{pmatrix} E & 0 \\ 0 & E \end{pmatrix} + Det\begin{pmatrix} 0 & -m \\ -m & 0 \end{pmatrix} + Det\begin{pmatrix} 0 & -i|\mathbf{p}| \\ i|\mathbf{p}| & 0 \end{pmatrix} =$$

Journal of Consciousness Exploration & Research | December 2010 | Vol. 1 | Issue 9 | pp. 80-109

Hu, H. &Wu, M. *The Principle of Existence II: Genesis of Self-Referential Matrix Law, & the Ontology & Mathematics of Ether*

92

$$Det\left(\begin{pmatrix} E & 0 \\ 0 & E \end{pmatrix} + \begin{pmatrix} 0 & -m \\ -m & 0 \end{pmatrix} + \begin{pmatrix} 0 & -i|\mathbf{p}| \\ i|\mathbf{p}| & 0 \end{pmatrix}\right) = Det\begin{pmatrix} E & -m-i|\mathbf{p}| \\ -m+i|\mathbf{p}| & E \end{pmatrix} = Det(L_M) \quad (3.43)$$

3.5 Hierarchical Natural Laws

The Natural laws created in accordance with the principle of existence are hierarchical and comprised of: (1) immanent Law of Conservation manifesting and governing in the external or internal world which may be violated in certain processes; (2) immanent Law of Zero manifesting and governing in the dual world as a whole; and (3) transcendental Law of One manifesting and governing in prespacetime. By ways of examples, conservations of energy, momentum and mass are immanent (and approximate) laws manifesting and governing in the external or internal world. Conservations of energy, momentums or mass to zero in the dual world comprised of the external world and internal world are immanent law manifesting and governing in the dual world as a whole. Conservation of One (Unity) based on Energy-Momentum- Mass Relationship is transcendental law manifesting and governing in prespacetime which is the foundation of external world and internal world.

4. SCIENTIFIC GENESIS OF ELEMENTARY PARTICLES

In the beginning Consciousness Created External &
Internal Objects & the governing Matrix Law

4.1 Scientific Genesis of Primordial Entities (Elementary Particles)

In the principle of existence, Consciousness creates, sustains and causes evolution of a free plane-wave fermion such as an electron in Dirac form as follows:

$$1 = e^{i0} = e^{i0}e^{i0} = e^{-iL+iL}e^{-iM+iM}$$

$$(\cos L - i \sin L)(\cos L + i \sin L)e^{-iM+iM} =$$

$$\left(\frac{m}{E} - i\frac{|\mathbf{p}|}{E}\right)\left(\frac{m}{E} + i\frac{|\mathbf{p}|}{E}\right)e^{-ip^\mu x_\mu + ip^\mu x_\mu} =$$

$$= \left(\frac{m-i|\mathbf{p}|}{E}\right)\left(\frac{m+i|\mathbf{p}|}{E}\right)e^{-ip^\mu x_\mu + ip^\mu x_\mu}$$

$$= \left(\frac{m^2 + \mathbf{p}^2}{E^2}\right)e^{-ip^\mu x_\mu + ip^\mu x_\mu} = \frac{E^2 - m^2}{\mathbf{p}^2}e^{-ip^\mu x_\mu + ip^\mu x_\mu}$$

$$= \left(\frac{E-m}{-|\mathbf{p}|}\right)\left(\frac{-|\mathbf{p}|}{E+m}\right)^{-1}\left(e^{-ip^{\mu}x_{\mu}}\right)\left(e^{-ip^{\mu}x_{\mu}}\right)^{-1} \rightarrow \qquad (4.1)$$

$$\frac{E-m}{-|\mathbf{p}|}e^{-ip^{\mu}x_{\mu}} = \frac{-|\mathbf{p}|}{E+m}e^{-ip^{\mu}x_{\mu}} \rightarrow \frac{E-m}{-|\mathbf{p}|}e^{-ip^{\mu}x_{\mu}} - \frac{-|\mathbf{p}|}{E+m}e^{-ip^{\mu}x_{\mu}} = 0$$

$$\rightarrow \begin{pmatrix} E-m & -|\mathbf{p}| \\ -|\mathbf{p}| & E+m \end{pmatrix}\begin{pmatrix} a_{e,+}e^{-ip^{\mu}x_{\mu}} \\ a_{i,-}e^{-ip^{\mu}x_{\mu}} \end{pmatrix} = \begin{pmatrix} L_{M,e} & L_{M,i} \end{pmatrix}\begin{pmatrix} \psi_{e,+} \\ \psi_{i,-} \end{pmatrix} = L_M\psi = 0$$

$$\rightarrow \begin{pmatrix} E-m & -\boldsymbol{\sigma}\cdot\mathbf{p} \\ -\boldsymbol{\sigma}\cdot\mathbf{p} & E+m \end{pmatrix}\begin{pmatrix} A_{e,+}e^{-ip^{\mu}x_{\mu}} \\ A_{i,-}e^{-ip^{\mu}x_{\mu}} \end{pmatrix} = \begin{pmatrix} L_{M,e} & L_{M,i} \end{pmatrix}\begin{pmatrix} \psi_{e,+} \\ \psi_{i,-} \end{pmatrix} = L_M\psi = 0$$

that is:

$$\begin{pmatrix} (E-m)\psi_{e,+} = \boldsymbol{\sigma}\cdot\mathbf{p}\,\psi_{i,-} \\ (E+m)\psi_{i,-} = \boldsymbol{\sigma}\cdot\mathbf{p}\,\psi_{e,+} \end{pmatrix} \text{ or } \begin{pmatrix} i\partial_t\psi_{e,+} - m\psi_{e,+} = -i\boldsymbol{\sigma}\cdot\nabla\psi_{i,-} \\ i\partial_t\psi_{i,-} + m\psi_{i,-} = -i\boldsymbol{\sigma}\cdot\nabla\psi_{e,+} \end{pmatrix} \qquad (4.2)$$

where substitutions $E \rightarrow i\partial_t$ and $\mathbf{p} \rightarrow -i\nabla$ have been made so that components of L_M

can act on external and internal wave functions.

In the principle of existence, Consciousness creates, sustains and causes evolution of a free plane-wave antifermion such as a positron in Dirac form as follows:

$$1 = e^{i0} = e^{i0}e^{i0} = e^{+iL-iL}e^{+iM-iM}$$

$$(\cos L + i\sin L)(\cos L - i\sin L)e^{+iM-iM} =$$

$$\left(\frac{m}{E} + i\frac{|\mathbf{p}|}{E}\right)\left(\frac{m}{E} - i\frac{|\mathbf{p}|}{E}\right)e^{+ip^{\mu}x_{\mu}-ip^{\mu}x_{\mu}}$$

$$= \left(\frac{m+i|\mathbf{p}|}{E}\right)\left(\frac{m-i|\mathbf{p}|}{E}\right)e^{+ip^{\mu}x_{\mu}-ip^{\mu}x_{\mu}}$$

$$= \left(\frac{m^2+\mathbf{p}^2}{E^2}\right)e^{+ip^{\mu}x_{\mu}-ip^{\mu}x_{\mu}} = \frac{E^2-m^2}{\mathbf{p}^2}e^{+ip^{\mu}x_{\mu}-ip^{\mu}x_{\mu}}$$

$$= \left(\frac{E-m}{-|\mathbf{p}|}\right)\left(\frac{-|\mathbf{p}|}{E+m}\right)^{-1}\left(e^{+ip^{\mu}x_{\mu}}\right)\left(e^{+ip^{\mu}x_{\mu}}\right)^{-1} \rightarrow \qquad (4.3)$$

$$\frac{E-m}{-|\mathbf{p}|}e^{+ip^{\mu}x_{\mu}} = \frac{-|\mathbf{p}|}{E+m}e^{+ip^{\mu}x_{\mu}} \rightarrow \frac{E-m}{-|\mathbf{p}|}e^{+ip^{\mu}x_{\mu}} - \frac{-|\mathbf{p}|}{E+m}e^{+ip^{\mu}x_{\mu}} = 0$$

Journal of Consciousness Exploration & Research | December 2010 | Vol. 1 | Issue 9 | pp. 80-109   94

Hu, H. &Wu, M. *The Principle of Existence II: Genesis of Self-Referential Matrix Law, & the Ontology & Mathematics of Ether*

$$\rightarrow \begin{pmatrix} E-m & -|\mathbf{p}| \\ -|\mathbf{p}| & E+m \end{pmatrix} \begin{pmatrix} a_{e,-}e^{+ip^{\mu}x_{\mu}} \\ a_{i,+}e^{+ip^{\mu}x_{\mu}} \end{pmatrix} = \begin{pmatrix} L_{M,e} & L_{M,i} \end{pmatrix} \begin{pmatrix} \psi_{e,-} \\ \psi_{i,+} \end{pmatrix} = L_M\psi = 0$$

$$\rightarrow \begin{pmatrix} E-m & -\boldsymbol{\sigma}\cdot\mathbf{p} \\ -\boldsymbol{\sigma}\cdot\mathbf{p} & E+m \end{pmatrix} \begin{pmatrix} A_{e,-}e^{+ip^{\mu}x_{\mu}} \\ A_{i,+}e^{+ip^{\mu}x_{\mu}} \end{pmatrix} = \begin{pmatrix} L_{M,e} & L_{M,i} \end{pmatrix} \begin{pmatrix} \psi_{e,-} \\ \psi_{i,+} \end{pmatrix} = L_M\psi = 0$$

Similarly, in the principle of existence, Consciousness creates, sustains and causes evolution of a free plane-wave fermion in Weyl (chiral) form as follows:

$$1 = e^{i0} = e^{i0}e^{i0} = e^{-iL+iL}e^{-iM+iM}$$

$$(\cos L - i\sin L)(\cos L + i\sin L)e^{-iM+iM} =$$

$$\left(\frac{m}{E} - i\frac{|\mathbf{p}|}{E}\right)\left(\frac{m}{E} + i\frac{|\mathbf{p}|}{E}\right)e^{-ip^{\mu}x_{\mu}+ip^{\mu}x_{\mu}}$$

$$= \left(\frac{m-i|\mathbf{p}|}{E}\right)\left(\frac{m+i|\mathbf{p}|}{E}\right)e^{-ip^{\mu}x_{\mu}+ip^{\mu}x_{\mu}}$$

$$= \left(\frac{m^2+\mathbf{p}^2}{E^2}\right)e^{-ip^{\mu}x_{\mu}+ip^{\mu}x_{\mu}} = \frac{E^2-\mathbf{p}^2}{m^2}e^{-ip^{\mu}x_{\mu}+ip^{\mu}x_{\mu}}$$

$$= \left(\frac{E-|\mathbf{p}|}{-m}\right)\left(\frac{-m}{E+|\mathbf{p}|}\right)^{-1}\left(e^{-ip^{\mu}x_{\mu}}\right)\left(e^{-ip^{\mu}x_{\mu}}\right)^{-1} \rightarrow \qquad (4.4)$$

$$\frac{E-|\mathbf{p}|}{-m}e^{-ip^{\mu}x_{\mu}} = \frac{-m}{E+|\mathbf{p}|}e^{-ip^{\mu}x_{\mu}} \rightarrow \frac{E-|\mathbf{p}|}{-m}e^{-ip^{\mu}x_{\mu}} - \frac{-m}{E+|\mathbf{p}|}e^{-ip^{\mu}x_{\mu}} = 0$$

$$\rightarrow \begin{pmatrix} E-|\mathbf{p}| & -m \\ -m & E+|\mathbf{p}| \end{pmatrix} \begin{pmatrix} a_{e,l}e^{-ip^{\mu}x_{\mu}} \\ a_{i,r}e^{-ip^{\mu}x_{\mu}} \end{pmatrix} = \begin{pmatrix} L_{M,e} & L_{M,i} \end{pmatrix} \begin{pmatrix} \psi_{e,l} \\ \psi_{i,r} \end{pmatrix} = L_M\psi = 0$$

$$\rightarrow \begin{pmatrix} E-\boldsymbol{\sigma}\cdot\mathbf{p} & -m \\ -m & E+\boldsymbol{\sigma}\cdot\mathbf{p} \end{pmatrix} \begin{pmatrix} A_{e,l}e^{-ip^{\mu}x_{\mu}} \\ A_{i,r}e^{-ip^{\mu}x_{\mu}} \end{pmatrix} = \begin{pmatrix} L_{M,e} & L_{M,i} \end{pmatrix} \begin{pmatrix} \psi_{e,l} \\ \psi_{i,r} \end{pmatrix} = L_M\psi = 0$$

that is:

$$\begin{pmatrix} (E-\boldsymbol{\sigma}\cdot\mathbf{p})\psi_{e,l} = m\psi_{i,r} \\ (E+\boldsymbol{\sigma}\cdot\mathbf{p})\psi_{i,r} = m\psi_{e,l} \end{pmatrix} \quad \text{or} \quad \begin{pmatrix} i\partial_t\psi_{e,l} + i\boldsymbol{\sigma}\cdot\nabla\psi_{e,l} = m\psi_{i,r} \\ i\partial_t\psi_{i,r} - i\boldsymbol{\sigma}\cdot\nabla\psi_{i,-} = m\psi_{e,l} \end{pmatrix} \qquad (4.5)$$

In the principle of existence, Consciousness creates, sustains and causes evolution of a free plane-wave fermion in another form as follows:

$$1 = e^{i0} = e^{i0}e^{i0} = e^{-iL+iL}e^{-iM+iM}$$

$$(\cos L - i\sin L)(\cos L + i\sin L)e^{-iM+iM} =$$

$$\left(\frac{m}{E} - i\frac{|\mathbf{p}|}{E}\right)\left(\frac{m}{E} + i\frac{|\mathbf{p}|}{E}\right)e^{-ip^\mu x_\mu + ip^\mu x_\mu}$$

$$= \left(\frac{m - i|\mathbf{p}|}{E}\right)\left(\frac{m + i|\mathbf{p}|}{E}\right)e^{-ip^\mu x_\mu + ip^\mu x_\mu} \tag{4.6}$$

$$= \left(\frac{E}{-m+i\varepsilon|\mathbf{p}|}\right)\left(\frac{-m-i|\mathbf{p}|}{E}\right)^{-1}\left(e^{-ip^\mu x_\mu}\right)\left(e^{-ip^\mu x_\mu}\right)^{-1}$$

$$\rightarrow \frac{E}{-m+i|\mathbf{p}|}e^{-ip^\mu x_\mu} = \frac{-m-i|\mathbf{p}|}{E}e^{-ip^\mu x_\mu}$$

$$\rightarrow \frac{E}{-m+i|\mathbf{p}|}e^{-ip^\mu x_\mu} - \frac{-m-i|\mathbf{p}|}{E}e^{-ip^\mu x_\mu} = 0$$

$$\rightarrow \begin{pmatrix} E & -m-i|\mathbf{p}| \\ -m+i|\mathbf{p}| & E \end{pmatrix}\begin{pmatrix} a_e e^{-ip^\mu x_\mu} \\ a_i e^{-ip^\mu x_\mu} \end{pmatrix} = \begin{pmatrix} L_{M,e} & L_{M,i} \end{pmatrix}\begin{pmatrix} \psi_e \\ \psi_i \end{pmatrix} = L_M\psi = 0$$

$$\rightarrow \begin{pmatrix} E & -m-i\boldsymbol{\sigma}\cdot\mathbf{p} \\ -m+i\boldsymbol{\sigma}\cdot\mathbf{p} & E \end{pmatrix}\begin{pmatrix} A_e e^{-ip^\mu x_\mu} \\ A_i e^{-ip^\mu x_\mu} \end{pmatrix} = \begin{pmatrix} L_{M,e} & L_{M,i} \end{pmatrix}\begin{pmatrix} \psi_e \\ \psi_i \end{pmatrix} = L_M\psi = 0$$

$$\rightarrow \begin{pmatrix} E & -Q \\ -Q^* & E \end{pmatrix}\begin{pmatrix} A_e e^{-ip^\mu x_\mu} \\ A_i e^{-ip^\mu x_\mu} \end{pmatrix} = \begin{pmatrix} L_{M,e} & L_{M,i} \end{pmatrix}\begin{pmatrix} \psi_e \\ \psi_i \end{pmatrix} = L_M\psi = 0$$

(where $Q = m + i\boldsymbol{\sigma}\cdot\mathbf{p}$ is a quaternion and $Q^* = m - i\boldsymbol{\sigma}\cdot\mathbf{p}$ is its conjugate) that is:

$$\begin{pmatrix} E\psi_e = (m + i\boldsymbol{\sigma}\cdot\mathbf{p})\psi_i \\ E\psi_i = (m - i\boldsymbol{\sigma}\cdot\mathbf{p})\psi_e \end{pmatrix} \quad \text{or} \quad \begin{pmatrix} i\partial_t\psi_e = m\psi_i + \boldsymbol{\sigma}\cdot\nabla\psi_i \\ i\partial_t\psi_i = m\psi_e - \boldsymbol{\sigma}\cdot\nabla\psi_i \end{pmatrix} \tag{4.7}$$

In the principle of existence, Consciousness creates, sustains and causes evolution of a linear plane-wave photon as follows:

$$1 = e^{i0} = e^{i0}e^{i0} = e^{-iL+iL}e^{-iM+iM}$$

$$(\cos L - i\sin L)(\cos L + i\sin L)e^{-iM+iM} =$$

$$\left(\frac{0}{E} - i\frac{|\mathbf{p}|}{E}\right)\left(\frac{0}{E} + i\frac{|\mathbf{p}|}{E}\right)e^{-ip^\mu x_\mu + ip^\mu x_\mu}$$

$$= \left(-i\frac{|\mathbf{p}|}{E}\right)\left(+i\frac{|\mathbf{p}|}{E}\right)e^{-ip^\mu x_\mu + ip^\mu x_\mu}$$

$$\left(\frac{\mathbf{p}^2}{E^2}\right)e^{-ip^\mu x_\mu + ip^\mu x_\mu} = \left(\frac{E^2}{\mathbf{p}^2}\right)e^{-ip^\mu x_\mu + ip^\mu x_\mu} =$$

$$\left(\frac{E}{-|\mathbf{p}|}\right)\left(\frac{-|\mathbf{p}|}{E}\right)^{-1}\left(e^{-ip^\mu x_\mu}\right)\left(e^{-ip^\mu x_\mu}\right)^{-1} \to$$

$$\frac{E}{-|\mathbf{p}|}e^{-ip^\mu x_\mu} = \frac{-|\mathbf{p}|}{E}e^{-ip^\mu x_\mu} \to \frac{E}{-|\mathbf{p}|}e^{-ip^\mu x_\mu} - \frac{-|\mathbf{p}|}{E}e^{-ip^\mu x_\mu} = 0$$

$$\to \begin{pmatrix} E & -|\mathbf{p}| \\ -|\mathbf{p}| & E \end{pmatrix}\begin{pmatrix} a_{e,+}e^{-ip^\mu x_\mu} \\ a_{i,-}e^{-ip^\mu x_\mu} \end{pmatrix} = \begin{pmatrix} L_{M,e} & L_{M,i} \end{pmatrix}\begin{pmatrix} \psi_{e,+} \\ \psi_{i,-} \end{pmatrix} = L_M\psi = 0$$

$$\to \begin{pmatrix} E & -\mathbf{s}\cdot\mathbf{p} \\ -\mathbf{s}\cdot\mathbf{p} & E \end{pmatrix}\begin{pmatrix} E_{0e,+}e^{-ip^\mu x_\mu} \\ iB_{0i,-}e^{-ip^\mu x_\mu} \end{pmatrix} = \begin{pmatrix} L_{M,e} & L_{M,i} \end{pmatrix}\begin{pmatrix} \psi_{e,+} \\ \psi_{i,-} \end{pmatrix} = L_M\psi_{photon} = 0$$

(4.8)

This photon wave function can be written as:

$$\psi_{photon} = \begin{pmatrix} \psi_{e,+} \\ \psi_{i,-} \end{pmatrix} = \begin{pmatrix} \mathbf{E} \\ i\mathbf{B} \end{pmatrix} = \begin{pmatrix} \mathbf{E}_0 e^{-i(\omega t - \mathbf{k}\cdot\mathbf{x})} \\ i\mathbf{B}_0 e^{-i(\omega t - \mathbf{k}\cdot\mathbf{x})} \end{pmatrix} = \begin{pmatrix} \mathbf{E}_0 \\ i\mathbf{B}_0 \end{pmatrix} e^{-i(\omega t - \mathbf{k}\cdot\mathbf{x})}$$ (4.9)

After the substitutions $E \to i\partial_t$ and $\mathbf{p} \to -i\nabla$, we have from the last expression in (4.8):

$$\begin{pmatrix} i\partial_t & i\mathbf{s}\cdot\nabla \\ i\mathbf{s}\cdot\nabla & i\partial_t \end{pmatrix}\begin{pmatrix} \mathbf{E} \\ i\mathbf{B} \end{pmatrix} = 0 \to \begin{pmatrix} \partial_t\mathbf{E} = \nabla\times\mathbf{B} \\ \partial_t\mathbf{B} = -\nabla\times\mathbf{E} \end{pmatrix}$$ (4.10)

where we have used the relationship $\mathbf{S}\cdot(-i\nabla) = \nabla\times$ to derive the latter equations which together with $\nabla\cdot\mathbf{E} = \mathbf{0}$ and $\nabla\cdot\mathbf{B} = \mathbf{0}$ are the Maxwell equations in the source-free vacuum.

In the principle of existence, Consciousness creates a neutrino in Dirac form, if Consciousness does, by replacing the last step of expression (3.87) with the following:

$$\rightarrow \begin{pmatrix} E & -\boldsymbol{\sigma} \cdot \mathbf{p} \\ -\boldsymbol{\sigma} \cdot \mathbf{p} & E \end{pmatrix} \begin{pmatrix} a_{e,+} e^{-ip^\mu x_\mu} \\ a_{i,-} e^{-ip^\mu x_\mu} \end{pmatrix} = \begin{pmatrix} L_{M,e} & L_{M,i} \end{pmatrix} \begin{pmatrix} \psi_{e,+} \\ \psi_{i,-} \end{pmatrix} = L_M \psi = 0 \qquad (4.11)$$

In the principle of existence, Consciousness creates, sustains and causes evolution of a linear plane-wave antiphoton as follows:

$$1 = e^{i0} = e^{i0} e^{i0} = e^{+iL - iL} e^{+iM - iM}$$

$$(\cos L + i \sin L)(\cos L - i \sin L) e^{+iM - iM} =$$

$$\left(\frac{0}{E} + i \frac{|\mathbf{p}|}{E} \right) \left(\frac{0}{E} - i \frac{|\mathbf{p}|}{E} \right) e^{+ip^\mu x_\mu - ip^\mu x_\mu}$$

$$= \left(+i \frac{|\mathbf{p}|}{E} \right) \left(-i \frac{|\mathbf{p}|}{E} \right) e^{+ip^\mu x_\mu - ip^\mu x_\mu}$$

$$\left(\frac{\mathbf{p}^2}{E^2} \right) e^{+ip^\mu x_\mu - ip^\mu x_\mu} = \left(\frac{E^2}{\mathbf{p}^2} \right) e^{+ip^\mu x_\mu - ip^\mu x_\mu} =$$

$$\left(\frac{E}{-|\mathbf{p}|} \right) \left(\frac{-|\mathbf{p}|}{E} \right)^{-1} \left(e^{+ip^\mu x_\mu} \right) \left(e^{+ip^\mu x_\mu} \right)^{-1} \rightarrow$$

$$\frac{E}{-|\mathbf{p}|} e^{+ip^\mu x_\mu} = \frac{-|\mathbf{p}|}{E} e^{+ip^\mu x_\mu} \rightarrow \frac{E}{-|\mathbf{p}|} e^{+ip^\mu x_\mu} - \frac{-|\mathbf{p}|}{E} e^{+ip^\mu x_\mu} = 0 \qquad (4.12)$$

$$\rightarrow \begin{pmatrix} E & -|\mathbf{p}| \\ -|\mathbf{p}| & E \end{pmatrix} \begin{pmatrix} \psi_{e,-} \\ \psi_{i,+} \end{pmatrix} = \begin{pmatrix} L_{M,e} & L_{M,i} \end{pmatrix} \begin{pmatrix} \psi_{e,-} \\ \psi_{i,+} \end{pmatrix} = L_M \psi = 0$$

$$\rightarrow \begin{pmatrix} E & -\mathbf{s} \cdot \mathbf{p} \\ -\mathbf{s} \cdot \mathbf{p} & E \end{pmatrix} \begin{pmatrix} i\mathrm{B}_{0e,-} e^{+ip^\mu x_\mu} \\ \mathrm{E}_{0i,+} e^{+ip^\mu x_\mu} \end{pmatrix} = \begin{pmatrix} L_{M,e} & L_{M,i} \end{pmatrix} \begin{pmatrix} \psi_{e,-} \\ \psi_{i,+} \end{pmatrix} = L_M \psi_{antiphoton} = 0$$

This antiphoton wave function can also be written as:

$$\psi_{antiphoton} = \begin{pmatrix} \psi_{e,-} \\ \psi_{i,+} \end{pmatrix} = \begin{pmatrix} i\mathbf{B} \\ \mathbf{E} \end{pmatrix} = \begin{pmatrix} i\mathbf{B}_0 e^{i(\omega t - \mathbf{k} \cdot \mathbf{x})} \\ \mathbf{E}_0 e^{i(\omega t - \mathbf{k} \cdot \mathbf{x})} \end{pmatrix} = \begin{pmatrix} i\mathbf{B}_0 \\ \mathbf{E}_0 \end{pmatrix} e^{i(\omega t - \mathbf{k} \cdot \mathbf{x})} \qquad (4.13)$$

In the principle of existence, Consciousness creates an antineutrino in Dirac form, if Consciousness does, by replacing the last step of expression (4.12) with the following:

$$\rightarrow \begin{pmatrix} E & -\boldsymbol{\sigma}\cdot\mathbf{p} \\ -\boldsymbol{\sigma}\cdot\mathbf{p} & E \end{pmatrix}\begin{pmatrix} a_{e,-}e^{+ip^{\mu}x_{\mu}} \\ a_{i,+}e^{+ip^{\mu}x_{\mu}} \end{pmatrix} = \begin{pmatrix} L_{M,e} & L_{M,i} \end{pmatrix}\begin{pmatrix} \psi_{e,-} \\ \psi_{i,+} \end{pmatrix} = L_M \psi = 0 \qquad (4.14)$$

Similarly, Consciousness likely creates and sustains spaceless (space/distance independent) external and internal wave functions of a mass m in Weyl (chiral) form as follows:

$$1 = e^{i0} = e^{i0}e^{i0} = e^{-iL+iL}e^{-iM+iM}$$

$$(\cos L - i\sin L)(\cos L + i\sin L)e^{-iM+iM} =$$

$$\left(\frac{m}{E} - i\frac{0}{E}\right)\left(\frac{m}{E} + i\frac{0}{E}\right)e^{-imt+imt} =$$

$$= \left(\frac{m}{E}\right)\left(\frac{m}{E}\right)e^{-imt+imt}$$

$$\left(\frac{m^2}{E^2}\right)e^{-imt+imt} = \left(\frac{E^2}{m^2}\right)e^{-imt+imt} =$$

$$\left(\frac{E}{-m}\right)\left(\frac{-m}{E}\right)^{-1}\left(e^{-imt}\right)\left(e^{-imt}\right)^{-1} \rightarrow$$

$$\frac{E}{-m}e^{-imt} = \frac{-m}{E}e^{-imt} \rightarrow \frac{E}{-m}e^{-imt} - \frac{-m}{E}e^{-imt} = 0 \qquad (4.15)$$

$$\rightarrow \begin{pmatrix} E & -m \\ -m & E \end{pmatrix}\begin{pmatrix} g_{W,e}e^{-imt} \\ g_{W,i}e^{-imt} \end{pmatrix} = \begin{pmatrix} L_{M,e} & L_{M,i} \end{pmatrix}\begin{pmatrix} V_{W,e} \\ V_{W,i} \end{pmatrix} = L_M V_W = 0$$

In the principle of existence, Consciousness likely creates, sustains and causes evolution of a spatially self-confined entity such as a proton through imaginary momentum $\mathbf{p}_i$ (downward self-reference such that $m^2 > E^2$) in Dirac form as follows:

$$1 = e^{i0} = e^{i0}e^{i0} = e^{+iL-iL}e^{+iM-iM}$$

$$(\cos L + i\sin L)(\cos L - i\sin L)e^{+iM-iM} =$$

$$\left(\frac{m}{E} + i\frac{|\mathbf{p}_i|}{E}\right)\left(\frac{m}{E} - i\frac{|\mathbf{p}_i|}{E}\right)e^{+ip^\mu x_\mu - ip^\mu x_\mu}$$

$$= \left(\frac{m + i|\mathbf{p}_i|}{E}\right)\left(\frac{m - i|\mathbf{p}_i|}{E}\right)e^{+ip^\mu x_\mu - ip^\mu x_\mu}$$

$$= \left(\frac{m^2 + \mathbf{p}_i^2}{E^2}\right)e^{+ip^\mu x_\mu - ip^\mu x_\mu} = \frac{E^2 - m^2}{\mathbf{p}_i^2}e^{+ip^\mu x_\mu - ip^\mu x_\mu}$$

$$= \left(\frac{E-m}{-|\mathbf{p}_i|}\right)\left(\frac{-|\mathbf{p}_i|}{E+m}\right)^{-1}\left(e^{+ip^\mu x_\mu}\right)\left(e^{+ip^\mu x_\mu}\right)^{-1} \rightarrow$$

$$\frac{E-m}{-|\mathbf{p}_i|}e^{+ip^\mu x_\mu} = \frac{-|\mathbf{p}_i|}{E+m}e^{+ip^\mu x_\mu} \rightarrow \frac{E-m}{-|\mathbf{p}_i|}e^{+ip^\mu x_\mu} - \frac{-|\mathbf{p}_i|}{E+m}e^{+ip^\mu x_\mu} = 0$$

$$\rightarrow \begin{pmatrix} E-m & -|\mathbf{p}_i| \\ -|\mathbf{p}_i| & E+m \end{pmatrix}\begin{pmatrix} s_{e,-}e^{+iEt} \\ s_{i,+}e^{+iEt} \end{pmatrix} = \begin{pmatrix} L_{M,e} & L_{M,i} \end{pmatrix}\begin{pmatrix} \psi_{e,-} \\ \psi_{i,+} \end{pmatrix} = L_M\psi = 0 \qquad (4.16)$$

After spinization of the last expression in (4.16), we have:

$$\rightarrow \begin{pmatrix} E-m & -\boldsymbol{\sigma}\cdot\mathbf{p}_i \\ -\boldsymbol{\sigma}\cdot\mathbf{p}_i & E+m \end{pmatrix}\begin{pmatrix} S_{e,-}e^{+iEt} \\ S_{i,+}e^{+iEt} \end{pmatrix} = \begin{pmatrix} L_{M,e} & L_{M,i} \end{pmatrix}\begin{pmatrix} \psi_{e,-} \\ \psi_{i,+} \end{pmatrix} = L_M\psi = 0 \qquad (4.17)$$

As discussed previously, it is likely that the last expression in (4.16) governs the confinement structure of the unspinized proton in Dirac form through imaginary momentum $\mathbf{p}_i$ and, on the other hand, expression (4.17) governs the confinement structure of spinized proton through $\mathbf{p}_i$.

Thus, an unspinized and spinized antiproton in Dirac form may be respectively governed as follows:

$$\begin{pmatrix} E-m & -|\mathbf{p}_i| \\ -|\mathbf{p}_i| & E+m \end{pmatrix}\begin{pmatrix} s_{e,+}e^{-iEt} \\ s_{i,-}e^{-iEt} \end{pmatrix} = \begin{pmatrix} L_{M,e} & L_{M,i} \end{pmatrix}\begin{pmatrix} \psi_{D,e} \\ \psi_{D,i} \end{pmatrix} = L_M\psi_D = 0 \qquad (4.18)$$

$$\begin{pmatrix} E-m & -\boldsymbol{\sigma}\cdot\mathbf{p}_i \\ -\boldsymbol{\sigma}\cdot\mathbf{p}_i & E+m \end{pmatrix}\begin{pmatrix} S_{e,+}e^{-iEt} \\ S_{i,-}e^{-iEt} \end{pmatrix} = \begin{pmatrix} L_{M,e} & L_{M,i} \end{pmatrix}\begin{pmatrix} \psi_{D,e} \\ \psi_{D,i} \end{pmatrix} = L_M\psi_D = 0 \qquad (4.19)$$

Similarly, in the principle of existence, Consciousness likely creates, sustains and causes evolution of a spatially self-confined entity such as a proton through imaginary momentum

$\mathbf{p}_i$ (downward self-reference) in Weyl (chiral) form as follows:

$$1 = e^{i0} = e^{i0}e^{i0} = e^{+iL-iL}e^{+iM-iM}$$

$$\left(\cos L + i\sin L\right)\left(\cos L - i\sin L\right)e^{+iM-iM} =$$

$$\left(\frac{m}{E} + i\frac{|\mathbf{p}_i|}{E}\right)\left(\frac{m}{E} - i\frac{|\mathbf{p}_i|}{E}\right)e^{+ip^\mu x_\mu - ip^\mu x_\mu}$$

$$= \left(\frac{m + i|\mathbf{p}_i|}{E}\right)\left(\frac{m - i|\mathbf{p}_i|}{E}\right)e^{+ip^\mu x_\mu - ip^\mu x_\mu}$$

$$= \left(\frac{m^2 + \mathbf{p}_i^2}{E^2}\right)e^{+ip^\mu x_\mu - ip^\mu x_\mu} = \frac{E^2 - \mathbf{p}_i^2}{m^2}e^{+ip^\mu x_\mu - ip^\mu x_\mu} =$$

$$\left(\frac{E - |\mathbf{p}_i|}{-m}\right)\left(\frac{-m}{E + |\mathbf{p}_i|}\right)^{-1}\left(e^{+ip^\mu x_\mu}\right)\left(e^{+ip^\mu x_\mu}\right)^{-1} \rightarrow$$

$$\frac{E - |\mathbf{p}_i|}{-m}e^{+ip^\mu x_\mu} = \frac{-m}{E + |\mathbf{p}_i|}e^{+ip^\mu x_\mu} \rightarrow \frac{E - |\mathbf{p}_i|}{-m}e^{+ip^\mu x_\mu} - \frac{-m}{E + |\mathbf{p}_i|}e^{+ip^\mu x_\mu} = 0$$

$$\rightarrow \begin{pmatrix} E - |\mathbf{p}_i| & -m \\ -m & E + |\mathbf{p}_i| \end{pmatrix}\begin{pmatrix} s_{e,r}e^{+iEt} \\ s_{i,l}e^{+iEt} \end{pmatrix} = \begin{pmatrix} L_{M,e} & L_{M,i} \end{pmatrix}\begin{pmatrix} \psi_{e,r} \\ \psi_{i,l} \end{pmatrix} = L_M\psi = 0 \quad (4.20)$$

After spinization of expression (3.114), we have:

$$\rightarrow \begin{pmatrix} E - \boldsymbol{\sigma}\cdot\mathbf{p}_i & -m \\ -m & E + \boldsymbol{\sigma}\cdot\mathbf{p}_i \end{pmatrix}\begin{pmatrix} S_{e,r}e^{+iEt} \\ S_{i,l}e^{+iEt} \end{pmatrix} = \begin{pmatrix} L_{M,e} & L_{M,i} \end{pmatrix}\begin{pmatrix} \psi_{e,r} \\ \psi_{i,l} \end{pmatrix} = L_M\psi = 0 \quad (4.21)$$

It is likely that the last expression in (4.20) governs the structure of the unspinized proton in Weyl form and expression (4.21) governs the structure of spinized proton in Weyl form.

Thus, an unspinized and spinized antiproton in Weyl form may be respectively governed as follows:

$$\begin{pmatrix} E - |\mathbf{p}_i| & -m \\ -m & E + |\mathbf{p}_i| \end{pmatrix}\begin{pmatrix} s_{e,l}e^{-iEt} \\ s_{i,r}e^{-iEt} \end{pmatrix} = \begin{pmatrix} L_{M,e} & L_{M,i} \end{pmatrix}\begin{pmatrix} \psi_{e,l} \\ \psi_{i,r} \end{pmatrix} = L_M\psi = 0 \quad (4.22)$$

$$\begin{pmatrix} E - \boldsymbol{\sigma}\cdot\mathbf{p}_i & -m \\ -m & E + \boldsymbol{\sigma}\cdot\mathbf{p}_i \end{pmatrix}\begin{pmatrix} S_{e,l}e^{-iEt} \\ S_{i,r}e^{-iEt} \end{pmatrix} = \begin{pmatrix} L_{M,e} & L_{M,i} \end{pmatrix}\begin{pmatrix} \psi_{e,l} \\ \psi_{i,r} \end{pmatrix} = L_M\psi = 0 \quad (4.23)$$

4.2 Scientific Genesis of Composite Entities

Then, in the principle of existence, Consciousness may create, sustain and cause evolution of a neutron in Dirac form which is comprised of an unspinized proton:

$$\left(\begin{pmatrix} E-e\phi-m & -|\mathbf{p}_i-e\mathbf{A}| \\ -|\mathbf{p}_i-e\mathbf{A}| & E-e\phi+m \end{pmatrix}\begin{pmatrix} s_{e,-}e^{+iEt} \\ s_{i,+}e^{+iEt} \end{pmatrix}=0\right)_p \tag{4.24}$$

and a spinized electron:

$$\left(\begin{pmatrix} E+e\phi-V-m & -\boldsymbol{\sigma}\cdot(\mathbf{p}+e\mathbf{A}) \\ -\boldsymbol{\sigma}\cdot(\mathbf{p}+e\mathbf{A}) & E+e\phi-V+m \end{pmatrix}\begin{pmatrix} S_{e,+}e^{-iEt} \\ S_{i,-}e^{-iEt} \end{pmatrix}=0\right)_e \tag{4.25}$$

as follows:

$$1 = e^{i0} = e^{i0}e^{i0}e^{i0}e^{i0} = \left(e^{i0}e^{i0}\right)_p\left(e^{i0}e^{i0}\right)_e = \left(e^{+iL-iM}e^{+iM-iM}\right)_p\left(e^{-iL+iL}e^{-iM+iM}\right)_e$$

$$= \left((\cos L + i\sin L)(\cos L - i\sin L)e^{+iM-iM}\right)_p\left((\cos L - i\sin L)(\cos L + i\sin L)e^{-iM+iM}\right)_e$$

$$= \left(\left(\frac{m}{E}+i\frac{|\mathbf{p}_i|}{E}\right)\left(\frac{m}{E}-i\frac{|\mathbf{p}_i|}{E}\right)e^{+ip^\mu x_\mu-ip^\mu x_\mu}\right)_p\left(\left(\frac{m}{E}-i\frac{|\mathbf{p}|}{E}\right)\left(\frac{m}{E}+i\frac{|\mathbf{p}|}{E}\right)e^{-ip^\mu x_\mu+ip^\mu x_\mu}\right)_e$$

$$= \left(\frac{m^2+\mathbf{p}_i^2}{E^2}e^{+ip^\mu x_\mu-ip^\mu x_\mu}\right)_p\left(\frac{m^2+\mathbf{p}^2}{E^2}e^{-ip^\mu x_\mu+ip^\mu x_\mu}\right)_e$$

$$= \left(\frac{E^2-m^2}{\mathbf{p}_i^2}e^{+ip^\mu x_\mu-ip^\mu x_\mu}\right)_p\left(\frac{E^2-m^2}{\mathbf{p}^2}e^{-ip^\mu x_\mu+ip^\mu x_\mu}\right)_e =$$

$$\left(\begin{pmatrix} E-m \\ -|\mathbf{p}_i| \end{pmatrix}\begin{pmatrix} -|\mathbf{p}_i| \\ E+m \end{pmatrix}^{-1}\left(e^{+ip^\mu x_\mu}\right)\left(e^{+ip^\mu x_\mu}\right)^{-1}\right)_p\left(\begin{pmatrix} E-m \\ -|\mathbf{p}| \end{pmatrix}\begin{pmatrix} -|\mathbf{p}| \\ E+m \end{pmatrix}^{-1}\left(e^{-ip^\mu x_\mu}\right)\left(e^{-ip^\mu x_\mu}\right)^{-1}\right)_e$$

$$\rightarrow \left(\begin{pmatrix} E-m & -|\mathbf{p}_i| \\ -|\mathbf{p}_i| & E+m \end{pmatrix}\begin{pmatrix} s_{e,-}e^{+iEt} \\ s_{i,+}e^{+iEt} \end{pmatrix}=0\right)_p\left(\begin{pmatrix} E-m & -|\mathbf{p}| \\ -|\mathbf{p}| & E+m \end{pmatrix}\begin{pmatrix} s_{e,+}e^{-iEt} \\ s_{i,-}e^{-iEt} \end{pmatrix}=0\right)_e$$

$$\rightarrow \left(\begin{array}{c} \left(\begin{pmatrix} E-e\phi-m & -|\mathbf{p}_i-e\mathbf{A}| \\ -|\mathbf{p}_i-e\mathbf{A}| & E-e\phi+m \end{pmatrix}\begin{pmatrix} s_{e,-}e^{+iEt} \\ s_{i,+}e^{+iEt} \end{pmatrix}=0\right)_p \\ \\ \left(\begin{pmatrix} E+e\phi-V-m & -\boldsymbol{\sigma}\cdot(\mathbf{p}+e\mathbf{A}) \\ -\boldsymbol{\sigma}\cdot(\mathbf{p}+e\mathbf{A}) & E+e\phi-V+m \end{pmatrix}\begin{pmatrix} S_{e,+}e^{-iEt} \\ S_{i,-}e^{-iEt} \end{pmatrix}=0\right)_e \end{array}\right)_n \tag{4.26}$$

In expressions (4.24), (4.25) and (4.26),$(\ )_p$, $(\ )_e$ and $(\ )_n$ indicate proton, electron and neutron respectively. Further, unspinized proton has charge e, electron has charge −e, $(A^\mu = (\phi, \mathbf{A}))_p$ and $(A^\mu = (\phi, \mathbf{A}))_e$ are the electromagnetic potentials acting on unspinized proton and tightly bound spinized electron respectively, and $(V)_e$ is a binding potential from the unspinized proton acting on the spinized electron causing tight binding as discussed later.

If $(A^\mu = (\phi, \mathbf{A}))_p$ is negligible due to the fast motion of the tightly bound spinized electron, we have from the last expression in (4.26):

$$\rightarrow \left(\begin{array}{l} \left(\begin{pmatrix} E-m & -|\mathbf{p}_i| \\ -|\mathbf{p}_i| & E+m \end{pmatrix} \begin{pmatrix} s_{e,-}e^{+iEt} \\ s_{i,+}e^{+iEt} \end{pmatrix} = 0 \right)_p \\ \\ \left(\begin{pmatrix} E+e\phi-V-m & -\boldsymbol{\sigma}\cdot(\mathbf{p}+e\mathbf{A}) \\ -\boldsymbol{\sigma}\cdot(\mathbf{p}+e\mathbf{A}) & E+e\phi-V+m \end{pmatrix} \begin{pmatrix} S_{e,+}e^{-iEt} \\ S_{i,-}e^{-iEt} \end{pmatrix} = 0 \right)_e \end{array} \right)_n \tag{4.27}$$

Experimental data on charge distribution and *g*-factor of neutron seem to support a neutron comprising of an unspinized proton and a tightly bound spinized electron.

The Weyl (chiral) form of the last expression in (4.26) and expression (4.27) are respectively as follows:

$$\left(\begin{array}{l} \left(\begin{pmatrix} -e\phi-|\mathbf{p}_i|-e\mathbf{A}| & -m \\ -m & -e\phi+|\mathbf{p}_i|-e\mathbf{A}| \end{pmatrix} \begin{pmatrix} s_{e,r}e^{+iEt} \\ s_{i,l}e^{+iEt} \end{pmatrix} = 0 \right)_p \\ \\ \left(\begin{pmatrix} E+e\phi-V-\boldsymbol{\sigma}\cdot(\mathbf{p}+e\mathbf{A}) & -m \\ -m & E+e\phi-V+\boldsymbol{\sigma}\cdot(\mathbf{p}+e\mathbf{A}) \end{pmatrix} \begin{pmatrix} S_{e,l}e^{-iEt} \\ S_{i,r}e^{-iEt} \end{pmatrix} = 0 \right)_e \end{array} \right)_n \tag{4.28}$$

$$\left(\begin{array}{l} \left(\begin{pmatrix} E-|\mathbf{p}_i| & -m \\ -m & E+|\mathbf{p}_i| \end{pmatrix} \begin{pmatrix} s_{e,r}e^{+iEt} \\ s_{i,l}e^{+iEt} \end{pmatrix} = 0 \right)_p \\ \\ \left(\begin{pmatrix} E+e\phi-V-\boldsymbol{\sigma}\cdot(\mathbf{p}+e\mathbf{A}) & -m \\ -m & E+e\phi-V+\boldsymbol{\sigma}\cdot(\mathbf{p}+e\mathbf{A}) \end{pmatrix} \begin{pmatrix} S_{e,l}e^{-iEt} \\ S_{i,r}e^{-iEt} \end{pmatrix} = 0 \right)_e \end{array} \right)_n \tag{4.29}$$

Then, in the principle of existence, Consciousness may create, sustain and cause evolution of a hydrogen atom comprising of a spinized proton:

Journal of Consciousness Exploration & Research | December 2010 | Vol. 1 | Issue 9 | pp. 80-109
Hu, H. &Wu, M. *The Principle of Existence II: Genesis of Self-Referential Matrix Law, & the Ontology & Mathematics of Ether*

103

$$\left(\begin{pmatrix} E-e\phi-m & -\boldsymbol{\sigma}\cdot(\mathbf{p}_i-e\mathbf{A}) \\ -\boldsymbol{\sigma}\cdot(\mathbf{p}_i-e\mathbf{A}) & E-e\phi+m \end{pmatrix} \begin{pmatrix} S_{e,-}e^{+iEt} \\ S_{i,+}e^{+iEt} \end{pmatrix} = 0 \right)_p \qquad (4.30)$$

and a spinized electron:

$$\left(\begin{pmatrix} E+e\phi-m & -\boldsymbol{\sigma}\cdot(\mathbf{p}+e\mathbf{A}) \\ -\boldsymbol{\sigma}\cdot(\mathbf{p}+e\mathbf{A}) & E+e\phi+m \end{pmatrix} \begin{pmatrix} S_{e,+}e^{-iEt} \\ S_{i,-}e^{-iEt} \end{pmatrix} = 0 \right)_e \qquad (4.31)$$

in Dirac form as follows:

$$1 = e^{i0} = e^{i0}e^{i0}e^{i0}e^{i0} = \left(e^{i0}e^{i0} \right)_p \left(e^{i0}e^{i0} \right)_e = \left(e^{+iL-iM}e^{+iM-iM} \right)_p \left(e^{-iL+iL}e^{-iM+iM} \right)_e$$

$$= \left((\cos L + i\sin L)(\cos L - i\sin L)e^{+iM-iM} \right)_p \left((\cos L - i\sin L)(\cos L + i\sin L)e^{-iM+iM} \right)_e$$

$$= \left(\left(\frac{m}{E} + i\frac{|\mathbf{p}_i|}{E} \right)\left(\frac{m}{E} - i\frac{|\mathbf{p}_i|}{E} \right)e^{+ip^\mu x_\mu - ip^\mu x_\mu} \right)_p \left(\left(\frac{m}{E} - i\frac{|\mathbf{p}|}{E} \right)\left(\frac{m}{E} + i\frac{|\mathbf{p}|}{E} \right)e^{-ip^\mu x_\mu + ip^\mu x_\mu} \right)_e$$

$$= \left(\frac{m^2 + \mathbf{p}_i^{\ 2}}{E^2}e^{+ip^\mu x_\mu - ip^\mu x_\mu} \right)_p \left(\frac{m^2 + \mathbf{p}^2}{E^2}e^{-ip^\mu x_\mu + ip^\mu x_\mu} \right)_e$$

$$= \left(\frac{E^2 - m^2}{\mathbf{p}_i^{\ 2}}e^{+ip^\mu x_\mu - ip^\mu x_\mu} \right)_p \left(\frac{E^2 - m^2}{\mathbf{p}^2}e^{-ip^\mu x_\mu + ip^\mu x_\mu} \right)_e =$$

$$\left(\left(\frac{E-m}{-|\mathbf{p}_i|} \right)\left(\frac{-|\mathbf{p}_i|}{E+m} \right)^{-1}\left(e^{+ip^\mu x_\mu} \right)\left(e^{+ip^\mu x_\mu} \right)^{-1} \right)_p \left(\left(\frac{E-m}{-|\mathbf{p}|} \right)\left(\frac{-|\mathbf{p}|}{E+m} \right)^{-1}\left(e^{-ip^\mu x_\mu} \right)\left(e^{-ip^\mu x_\mu} \right)^{-1} \right)_e$$

$$\rightarrow \left(\begin{pmatrix} E-m & -|\mathbf{p}_i| \\ -|\mathbf{p}_i| & E+m \end{pmatrix}\begin{pmatrix} s_{e,-}e^{+iEt} \\ s_{i,+}e^{+iEt} \end{pmatrix} = 0 \right)_p \left(\begin{pmatrix} E-m & -|\mathbf{p}| \\ -|\mathbf{p}| & E+m \end{pmatrix}\begin{pmatrix} s_{e,+}e^{-iEt} \\ s_{i,-}e^{-iEt} \end{pmatrix} = 0 \right)_e$$

$$\rightarrow \left(\begin{array}{l} \left(\begin{pmatrix} E-e\phi-m & -\boldsymbol{\sigma}\cdot(\mathbf{p}_i-e\mathrm{A}) \\ -\boldsymbol{\sigma}\cdot(\mathbf{p}_i-e\mathrm{A}) & E-e\phi+m \end{pmatrix}\begin{pmatrix} S_{e,-}e^{+iEt} \\ S_{i,+}e^{+iEt} \end{pmatrix} = 0 \right)_p \\ \left(\begin{pmatrix} E+e\phi-m & -\boldsymbol{\sigma}\cdot(\mathbf{p}+e\mathbf{A}) \\ -\boldsymbol{\sigma}\cdot(\mathbf{p}+e\mathbf{A}) & E+e\phi+m \end{pmatrix}\begin{pmatrix} S_{e,+}e^{-iEt} \\ S_{i,-}e^{-iEt} \end{pmatrix} = 0 \right)_e \end{array} \right)_h \qquad (4.32)$$

In expressions (4.30), (4.31) and (4.32), $(\ )_p$, $(\ )_e$ and $(\ )_h$ indicate proton, electron and hydrogen atom respectively. Again, proton has charge e, electron has charge −e, and

$\left(A^{\mu}=(\phi,\mathbf{A})\right)_{p}$ and $\left(A^{\mu}=(\phi,\mathbf{A})\right)_{e}$ are the electromagnetic potentials acting on spinized proton and spinized electron respectively.

Again, if $\left(A^{\mu}=(\phi,\mathbf{A})\right)_{p}$ is negligible due to fast motion of the orbiting spinized electron, we have from the last expression in (3.129):

$$\rightarrow \left(\begin{array}{l} \left(\begin{pmatrix} E-m & -\boldsymbol{\sigma}\cdot\mathbf{p}_i \\ -\boldsymbol{\sigma}\cdot\mathbf{p}_i & E+m \end{pmatrix}\begin{pmatrix} S_{e,-}e^{+iEt} \\ S_{i,+}e^{+iEt} \end{pmatrix}=0\right)_{p} \\ \left(\begin{pmatrix} E+e\phi-m & -\boldsymbol{\sigma}\cdot(\mathbf{p}+e\mathbf{A}) \\ -\boldsymbol{\sigma}\cdot(\mathbf{p}+e\mathbf{A}) & E+e\phi+m \end{pmatrix}\begin{pmatrix} S_{e,+}e^{-iEt} \\ S_{i,-}e^{-iEt} \end{pmatrix}=0\right)_{e} \end{array} \right)_{h} \quad (4.33)$$

The Weyl (chiral) form of the last expression in (3.129) and expression (3.130) are respectively as follows:

$$\left(\begin{array}{l} \left(\begin{pmatrix} E-e\phi-\boldsymbol{\sigma}\cdot(\mathbf{p}_i-e\mathbf{A}) & -m \\ -m & E-e\phi+\boldsymbol{\sigma}\cdot(\mathbf{p}_i-e\mathbf{A}) \end{pmatrix}\begin{pmatrix} S_{e,r}e^{+iEt} \\ S_{i,l}e^{+iEt} \end{pmatrix}=0\right)_{p} \\ \left(\begin{pmatrix} E+e\phi-\boldsymbol{\sigma}\cdot(\mathbf{p}+e\mathbf{A}) & -m \\ -m & E+e\phi+\boldsymbol{\sigma}\cdot(\mathbf{p}+e\mathbf{A}) \end{pmatrix}\begin{pmatrix} S_{e,l}e^{-iEt} \\ S_{i,r}e^{-iEt} \end{pmatrix}=0\right)_{e} \end{array} \right)_{h} \quad (4.34)$$

$$\left(\begin{array}{l} \left(\begin{pmatrix} E-\boldsymbol{\sigma}\cdot\mathbf{p}_i & -m \\ -m & E+\boldsymbol{\sigma}\cdot\mathbf{p}_i \end{pmatrix}\begin{pmatrix} S_{e,r}e^{+iEt} \\ S_{i,l}e^{+iEt} \end{pmatrix}=0\right)_{p} \\ \left(\begin{pmatrix} E+e\phi-\boldsymbol{\sigma}\cdot(\mathbf{p}+e\mathbf{A}) & -m \\ -m & E+e\phi+\boldsymbol{\sigma}\cdot(\mathbf{p}+e\mathbf{A}) \end{pmatrix}\begin{pmatrix} S_{e,l}e^{-iEt} \\ S_{i,r}e^{-iEt} \end{pmatrix}=0\right)_{e} \end{array} \right)_{h} \quad (4.35)$$

5. MATHEMATICS & ONTOLOGY OF ETHER

Ether is Mathematical,
Immanent & Transcendental

5.1 Mathematical Aspect of Ether

In the principle of existence, it is our comprehension that:

(1) The mathematical representation of the primordial ether in prespacetime is the Euler's number (Euler's Constant) e which makes the Euler's identity possible:

$$e^{i\pi} + 1 = 0 \qquad (5.1)$$

(2) Euler's number e is the foundation of primordial distinction in prespacetime:

$$1 = e^{i0} = e^{i0}e^{i0} = e^{iL-iL}e^{iM-iM} = e^{iL}e^{iM}e^{-iL}e^{-iM} = e^{-iL}e^{-iM}/e^{-iL}e^{-iM} = e^{iL}e^{iM}/e^{iL}e^{iM}\ldots \quad (5.2)$$

(3) Euler's number e is the foundation of the genesis of energy, momentum & mass relationship in prespacetime:

$$1 = e^{i0} = e^{-iL+iL} = L_e L_i^{-1} = (\cos L - i \sin L)(\cos L + i \sin L) = \qquad (5.3)$$

$$\left(\frac{m}{E} - i\frac{|\mathbf{p}|}{E}\right)\left(\frac{m}{E} + i\frac{|\mathbf{p}|}{E}\right) = \left(\frac{m - i|\mathbf{p}|}{E}\right)\left(\frac{m + i|\mathbf{p}|}{E}\right) = \left(\frac{m^2 + \mathbf{p}^2}{E^2}\right) \rightarrow$$

$$E^2 = m^2 + \mathbf{p}^2$$

(4) Euler's number e is the foundation of the genesis, sustenance and evolution of an elementary particle in prespacetime:

$$1 = e^{i0} = e^{i0}e^{i0} = e^{-iL+iL}e^{-iM+iM} = L_e L_i^{-1}\left(e^{-iM}\right)\left(e^{-iM}\right)^{-1} \rightarrow \qquad (5.4)$$

$$\left(L_{M,e} \quad L_{M,i}\right)\begin{pmatrix} A_e e^{-iM} \\ A_i e^{-iM} \end{pmatrix} = L_M \begin{pmatrix} A_e \\ A_i \end{pmatrix} e^{-iM} = L_M \begin{pmatrix} \psi_e \\ \psi_i \end{pmatrix} = L_M \psi = 0$$

(5) Euler's number e is also the foundation of quantum entanglement or gravity in prespacetime.

(6) Euler's number is immanent in the sense that it is the ingredient of (1) to (5) thus all "knowing" and all "present."

(7) Euler's number is also transcendental in the sense that is the foundation of existence thus "omnipotent" and beyond creation.

5.2 Immanent Aspect of Ether

In the principle of existence, the immanent aspect of ether associated with individual entity ("i-ether") has following attributes:

i-ether is the ingredient of atoms, of molecules, of cells, of a body;
i-ether is in space, time, motion, rest;
i-ether is governed by the laws of physics, chemistry, biology;
i-ether is the ingredient of this world, the Earth, the Solar System.

i-ether is the ingredient of awareness, feeling, imagination, free will;

Journal of Consciousness Exploration & Research | December 2010 | Vol. 1 | Issue 9 | pp. 80-109   106

Hu, H. &Wu, M. *The Principle of Existence II: Genesis of Self-Referential Matrix Law, & the Ontology & Mathematics of Ether*

i-ether is in love, passion, hope, despair;
i-ether is governed by the laws of psychology, economics, sociology;
i-ether is the ingredient of mind, soul, spirit.

In the principle of existence, the immanent of ether associated with the universal entity ("I-ETHER") has following attributes:

I-ETHER IS atoms, molecules, cells, body;
I-ETHER IS space, time, motion, rest;
I-ETHER IS laws of physics, chemistry, biology, physiology;
I-ETHER IS this World, the Earth, the Solar System.

I-ETHER IS awareness, feeling, imagination, free will;
I-ETHER IS love, passion, hope, despair;
I-ETHER IS the laws of psychology, economics, sociology;
I-ETHER IS mind, soul, spirit.

5.3 Transcendental Aspect of Ether

In the principle of existence, the transcendental aspect of ether associated with individual/entity ("t-ether") has following attributes:

t-ether is not the ingredient of atoms, of molecules, of cells, of a body;
t-ether is not in space, time, motion, rest;
t-ether is not governed by the laws of physics, chemistry, biology;
t-ether is not the ingredient of this world, the Earth, the Solar System.

t-ether is beyond awareness, feeling, imagination, free will;
t-ether is beyond love, passion, hope, despair;
t-ether is beyond the laws of psychology, economics, sociology;
t-ether is beyond mind, soul, spirit.

In the principle of existence, the transcendental aspect of ether associated with the universal entity ("T-ETHER") has following attributes:

T-ETHER IS NOT the atoms, molecules, cells, body;
T-ETHER IS NOT the space, time, motion, rest;
T-ETHER IS NOT the laws of physics, chemistry, biology;
T-ETHER IS NOT this world, the Earth, the Solar System.

T-ETHER IS NOT awareness, feeling, imagination, free will;
T-ETHER IS NOT love, passion, hope, despair;
T-ETHER IS NOT the laws of psychology, economics, sociology;
T-ETHER IS NOT mind, soul, spirit.

6. CONCLUSION

This work is the continuation of the principle of existence. It has mainly dealt with the genesis of self-referential Matrix Law and the ontology & mathematics of ether which have been discovered by us in continuation or rather revealed to us, the submitters to truth, by Consciousness. Yet again, we caution fellow truth seekers and dear readers that we as humans can only strive for perfection, completeness and correctness in our comprehensions and writings because we ourselves are limited and imperfect.

According to the principle of existence, in the beginning there was Consciousness (prespacetime) by itself $e^0 = 1$ materially empty and spiritually restless. And it began to imagine through primordial self-referential spin $1=e^{i0}=e^{i0}e^{i0}=e^{iL-iL}e^{iM-iM}=e^{iL}e^{iM}e^{-iL}e^{-iM}= e^{-iL}e^{-iM}/e^{-iL}e^{-iM}=e^{iL}e^{iM}/e^{iL}e^{iM}$ …such that it created the self-referential Matrix Law, the external object to be observed and internal object as observed, separated them into external world and internal world, caused them to interact through said Matrix Law and thus gave birth to the Universe which it has since passionately loved, sustained and made to evolve. The Natural laws created in accordance with the principle of existence are hierarchical and comprised of: (1) immanent Law of Conservation manifesting and governing in the external or internal world which may be violated in certain processes; (2) immanent Law of Zero manifesting and governing in the dual world as a whole; and (3) transcendental Law of One manifesting and governing in prespacetime.

Let it also be known that the principle of existence is supported by experiments (or has sound basis in empirical evidence), since experimentally, we demonstrated that: (1) Consciousness is associated with (or simply is) prespacetime and our brain is the vehicle for conscious experiences and interactions; and (2) there exists an instantaneous transcendental force (quantum entanglement or gravity) beyond spacetime which makes omnipotence, omnipresence and omniscience of Consciousness (prespacetime) possible and feasible (see Hu & Wu, 2001-2010).

In the principle of existence, the principles and mathematics which Consciousness have used to create, sustain and makes evolving of elementary particles are beautiful and simple. First, Consciousness employs the following ontological principles among others: (1) Principle of oneness/unity of existence through quantum entanglement in the body (ether) of prespacetime; and (2) Principle of hierarchical primordial self-referential spin creating:

- Energy-Momentum-Mass Relationship as Transcendental Law of One
- Energy-Momentum-Mass Relationship as Determinant of Matrix Law
- Dual-world Law of Zero of Energy, Momentum & Mass.
- Immanent Law of Conservation of Energy, Momentum & Mass   in External/Internal World which may be violated in certain processes.

Second, Consciousness employs the following mathematical elements & forms among others in order to empower the above ontological principles among others:

(1) *e*, Euler's number, for (to empower) ether (aether) as foundation/basis/medium of existence (body of prespacetime);

(2) *i*, imaginary number, for (to empower) thoughts and imagination;

(3) 0, zero, for (to empower) emptiness/undifferentiated/primordial state;

(4) 1, one, for (to empower) oneness/unity of existence;

(5) +, -, *, /, = for (to empower) creation, dynamics, balance & conservation;

(6) Pythagorean theorem for (to empower) Energy-Momentum-Mass Relationship; and

(7) *M*, matrix, for (to empower) the external and internal worlds (the Dual World) and the interaction of external and internal worlds.

DEDICATION:

We dedicate this work to Consciousness which created the self-referential Matrix Law, the external object to be observed and internal object as observed, separated them into external world and internal world, caused them to interact through said Matrix Law and thus gave birth to the Universe.

[SELF-]REFERENCE

Hu, H. & Wu, M. 2001a, Mechanism of anesthetic action: oxygen pathway perturbation hypothesis. Med. Hypotheses, 57: 619-627. Also see arXiv 2001b; physics/0101083.

Hu, H. & Wu, M. 2002, Spin-mediated consciousness theory. arXiv: quant-ph/0208068. Also see Med. Hypotheses 2004a: 63: 633-646.

Hu, H. & Wu, M. 2004b, Spin as primordial self-referential process driving quantum mechanics, spacetime dynamics and consciousness. NeuroQuantology, 2:41-49. Also see Cogprints: ID2827 2003.

Hu, H. & Wu, M. 2004c, Action potential modulation of neural spin networks suggests possible role of spin in memory and consciousness. NeuroQuantology, 2:309-316. Also see Cogprints: ID3458 2004d.

Hu, H. & Wu, M. 2006a, Thinking outside the box: the essence and implications of quantum entanglement. NeuroQuantology, 4: 5-16.

Hu, H. & Wu, M. 2006b, Photon induced non-local effect of general anesthetics on the brain. NeuroQuantology, 4: 17-31. Also see Progress in Physics 2006c; v3: 20-26.

Hu, H. & Wu, M. 2006d, Evidence of non-local physical, chemical and biological effects supports quantum brain. NeuroQuantology, 4: 291-306. Also see Progress in Physics 2007a; v2: 17-24.

Hu, H. & Wu, M. 2007b, Thinking outside the box II: the origin, implications and applications of gravity and its role in consciousness. NeuroQuantology, 5: 190-196.

Hu, H. & Wu, M. 2007c, On dark chemistry: what's dark matter and how mind influences brain through proactive spin. NeuroQuantology, 5: 205-213.

Hu, H. & Wu, M. 2008a, Concerning spin as mind-pixel: how mind interacts with the brain through electric spin effects. NeuroQuantology, 6: 26-31.

Hu, H. 2008b, The state of science, religion and consciousness. NeuroQuantology, 6: 323-332.

Hu, H. 2009, Quantum enigma - physics encounters consciousness (book review). Psyche, 15: 1-4.

Hu, H. & Wu, M. (2010), Let All Truth Seekers Be the Scientific & Spiritual Vessels to Carry Science & Religion to New Heights, Scientific GOD Journal 1:1, pp. 1-7.

Hu, H. & Wu, M. (2010), The Principle of Existence: Toward a Scientific Theory of Everything, Scientific GOD Journal 1:1, pp. 8-77.

Hu, H. (2010), GOD's Scientific Truth Is Marching on (Poem), Scientific GOD Journal 1:1, pp. 78-79.

Hu, H. & Wu, M. (2010), Let All Truth Seekers Be the Vessels to Carry Consciousness Research to New Heights, JCER 1:1, pp. 1-4.

Hu, H. & Wu, M. (2010),The Principle of Existence: Towards a Science of Consciousness, JCER 1:1, pp. 50-119.

Hu, H. (2010), Let All Truth Seekers Be the Vessels to Carry Physics Research to New Heights, Prespacetime journal 1:1, pp. 1-3.

Hu, H. & Wu, M. (2010), Prespacetime Model of Elementary Particles, Four Forces & Consciousness, Prespacetime journal 1:1, pp. 77-146.

Hu, H. (2010), Song to Immanence & Transcendence (Poem), Scientific GOD Journal 1:6, pp. 455-456.

Hu, H. (2010), Oh My Atheist Colleagues in Science (Poem), Scientific GOD Journal 1:7, pp. 518-519.

Hu, H. & Wu, M. (2010), Current Landscape and Future Direction of Theoretical & Experimental Quantum Brain/Mind/Consciousness Research, JCER 1:8, pp. 888-897.

Hu, H. & Wu, M. (2010), Experimental Support of Spin-mediated Consciousness Theory from Various Sources, JCER 1:8, pp. 907-936.

Hu, H. & Wu, M. (2010), Consciousness-mediated Spin Theory: The Transcendental Ground of Quantum Reality, JCER 1:8, pp. 937-970.

Journal of Consciousness Exploration & Research| December 2010 | Vol. 1 | Issue 9 | pp. 110-128

110

Hunter, M.D., Mulligan, B.P., Dotta, B. T., Saroka, K. S., Lavallee, C. F., Koren, S. A., & Persinger, M. A., *Cerebral Dynamics and Discrete Energy Changes in the Personal Physical Environment During Intuitive-Like States and Perceptions*

Article

Cerebral Dynamics and Discrete Energy Changes in the Personal Physical Environment During Intuitive-Like States and Perceptions

Mathew D. Hunter[1,2], Bryce P. Mulligan[1,2], Blake T. Dotta[1,2]
Kevin S. Saroka[1,3], Christina F. Lavallee[1,3], Stanley A. Koren[1,3],
Michael A. Persinger[1,2,3,4*]

Behavioural Neuroscience Laboratory[1]
Departments of Biology[2] and Psychology[3]
Behavioural Neuroscience, Human Studies, and Biomolecular Sciences Programs[4]
Laurentian University, Sudbury, Ontario, Canada

ABSTRACT

The attribution of unobservable cognitive states to others, a component of the "Theory of Mind", involves activity within the right temporoparietal region. We tested an exceptional subject, Sean Harribance, who displayed a reliable, consistent configuration of QEEG activity over this region that was confirmed through source localization software. The blind-rated accuracies of the histories of 40 people shown in 40 different photographs were strongly correlated with the quantitative occurrence of this conspicuous QEEG pattern displayed during Mr. Harribance's "intuitive state". The proportions of specific microstates were also strongly correlated with his accuracy of discerning the historical characteristics of the people in the photographs. Compared to the normal population his microstates were half the duration and his sense of "now" was about twice as fast as the average person. During his intuitive states there was strong congruence of activity between the left temporal lobes of participants who sat near Mr. Harribance and the activity over his temporal (primarily right) lobes within the theta and 19-20 Hz band. Reversible increases in photon emissions and small alterations in the intensity within the nearby (up to a 1 meter) geomagnetic field along the right side of his head were equivalent to energies of about 10^{-11} Joules with amplitude modulations in the 0.2 to 0.6 Hz range. The results indicate even exceptional skills previously attributed to aberrant sources are variations of normal cerebral dynamics associated with intuition and may involve small but discrete changes in proximal energy.

Key words: Theory of Mind; microstates; temporoparietal region; geomagnetic alterations; photon emissions; intuitive states.

* To whom correspondence should be addressed:

Michael Persinger
935 Ramsey Lake Road
Behavioural Neuroscience Laboratory
Departments of Biology and Psychology
Laurentian University
Sudbury ON, Canada
P3E 2C6
Telephone: 1-705-675-4824, or 1-705-675-1151 ext 4824
Fax Number: 1-705-671-3844
E-mail: mpersinger@laurentian.ca

1. Introduction

A fundamental human ability to predict and to interpret the behavior of others can be explained by a "Theory of Mind" (Happe 2003; Saxe and Wexler 2005). This is a process by which most healthy adults attribute unobservable cognitive states to others and integrate these states into a coherent model (Saxe and Wexler 2003; Vogeley 2001). Regions near the temporoparietal junction of the human brain have been implicated in a broad range of social cognitive tasks including recognition of faces and inferences about other's thoughts (Allison et al. 2000; Brunet et al. 2000; Fletcher et al. 1995; Hoffman and Haxby 2000).

Sean Harribance (SH) is a classic example of a person who possesses an ability to infer lifetime "experiences" of the putative memories of others while viewing photographs of a person or individuals known to this person while the latter is present. Most of his comments concerning the person are considered so exceptional, specific, or unique that his primary income for the last 40 years has been derived from these interpretations. His interpersonal behaviors are amicable and sincere but exhibit qualitatively different features that are clearly discernable by most people.

There has also been a long experimental and anecdotal history concerning SH and his capacity to "access" accurate information about others by mechanisms not known to date (Harribance 1994). Single photon emission computerized tomography (SPECT) displayed increased uptake of tracer and perfusion within the superior medial portion of the right parietal cortices during these experiences (Roll et al. 2002). Neuropsychological evidence suggested anomalous function within the frontoparietal-temporal region of the right hemisphere (Roll et al. 2002). During a previous visit (ten years ago) three different people, who were each given 10 pictures of their relatives and 10 comments (total of 30 photographs and 30 comments) by SH of the medical history and personal experiences of those people, correctly matched 8 out of 10 of the pictures with the descriptions.

By applying the modern tools of quantitative electroencephalography (QEEG) we found reliable signatures involving his right temporoparietal region that were associated with the ratings of accuracy for his statements. We also found consistent changes in QEEG activity within the right temporoparietal regions of the participants sitting near SH while he was "interpreting" the person or related photographs. We called this conspicuous, stable pattern the Harribance Configuration (HC). The overall pattern of activity and his subjective attributions suggested his experiences may be a variant of normal intuition (Kuo et al. 2009).

We have assumed that SH's unique ability to interpret such specific details of a person's life with no prior contact is primarily cerebrally generated and can be quantified by electroencephalography. On the bases of the proximity apparently required between SH and the person and the experienced physical changes reported by participants, we reasoned that the "information" might be obtained through a medium within the proximal environment. We hypothesized that two potential mediums may be a contributor to the reception of information: 1) biophoton emissions and 2) the shared geomagnetic field within which both SH and the subjects are immersed.

Biophotons have been shown to be potential neural communication signals (Sun et al, 2010).

According to Popp (1979) biological systems have the capacity to store coherent photons from the external world and emit a few hundred to about a thousand photons per cm^2 per second; he also hypothesized that photons may be utilized for cell-cell communication. The recent measurements that a photon is not massless (Liang-Cheng et al, 2005) have far reaching implications relevant to biological systems. They include deviations in the behaviour of static electromagnetic fields, longitudinal electromagnetic radiation and even questions of gravitational deflection (Liang-Cheng et al, 2005). Such small changes are even more important in light of the minute energies of about 10^{-20} J associated with an action potential (Persinger et al. 2008; Persinger, 2010) and the recent direct measurements that a single neuron can affect specific responses (Houweling and Brecht, 2008).

Biological systems emerged within the earth's magnetic field and human beings are immersed in the geomagnetic field. Fluctuations with peak-to-peak amplitudes of about 1% of the steady-state condition within the biofrequency to mHz range (Persinger, 1980) have powerful effects upon electroencephalographic activity (Babayev and Allahverdiyeva, 2007; Mulligan et al, 2010) and correlative behaviours that can be simulated experimentally (Michon and Persinger, 1997). That information could be stored within the space occupied by the earth's magnetic field has both theoretical and quantitative support (Persinger, 2009) although the definitive experiment to relate byte-dependent patterns to specific responses or ideas has not been completed. There is ample evidence that application of weak (microTesla to nanoTesla) physiologically-patterned magnetic fields, particularly over the right hemisphere, are associated with the report of common themes of experiences in normal volunteers (St-Pierre and Persinger, 2006, Persinger and Tiller 2008, Persinger et al. 2009).

2. Materials and Methods

2.1. Participants

All subjects who took part in this study did so with written informed consent. The procedures outlined were approved by Laurentian University's Research Ethics Board. Demographic information is given below with respect to each of the different paradigms. All testing was completed between 12:00-17:00 local time.

2.2. Data Acquisition and Analysis

2.2.1. Quantitative Electroencephalography (QEEG)

During his visit to our laboratory we measured SH over a period of five successive days, resulting in approximately 5 hours (500 to 1000 samples per sec) of QEEG measurements (8 or 19 channels) by different devices. Eight channels for F7, F8, T3, T4, P3, P4, O1, O2 from either of two Grass Instrument Model 8-16 C EEGs (16 channel) were monitored by hardcopy (paper). The filter selections for each channel were set for the standard range between 0.5 Hz and 35 Hz. Each Model 8-16C machine was interfaced via a custom shielded cable, a parallel analogue shield interface cable (Nat. Inst. SH100100) and a shielded connector block (Nat. Inst. SCB-100) to a National

Instruments PCI-607IE Multi I/O Board computer interface card. The data were extracted at 1000 Hz sampling (every 1 msec) by a DELL Dimension 8100 Personal Computer on a Windows 2000 Professional Platform. A custom designed user interface or Virtual Instrument (VI) using National Instruments Labview (Version 6.0i-2000) allowed the multichannel sample to be manually recorded to fixed disk.

A second QEEG utilized was a Mitsar 201 system amplifier which samples at 500 Hz with an input range of -500 to +500 microvolts and 16 bit analogue to digital conversion. The electrode cap (Electro-Cap International) utilized 19 AgCl electrodes using the 10-20 international standard method of electrode placement. Impedance for all electrodes was maintained at less than 10 kOhms. All electrodes were linked to ear references for monopolar measurements and appropriate channels for occasional bipolar recordings. WINEEG v2.82 was utilized for data collection, artifact removal, and spectral analyses. Visual inspection and independent component analysis were utilized for removal of all artifacts. Spectral analyses were computed utilizing WINEEG software and further statistical analysis of spectral components was completed using SPSS software. All EEG coherence results were completed utilizing EEG Lab software (Delorme and Makeig 2004).

Source localization was completed using sLORETA (standardized low resolution electromagnetic tomography; Jurcak et al. 2007; Pascual-Marqui 2002) software. sLORETA has cross-modal validation with respect to Brodmann area mapping with as few as 19 channels (Winterer et al. 2001, Mulert et al. 2004). Source localization analysis was utilized to assess how SH differed spectrally between his resting and IS ("interpretational state) conditions. The IS condition was completed without talking while he was "calling his angel". Four 1 minute sessions of IS were completed while wearing the 19 channel EEG from which eight 30 second artifact free EEG records were collected and compared with eight 30 second artifact free resting EEG sections.

2.2.2. Rated Accuracy of Types of Interpretations and EEG Patterns

The procedures preferred by SH were followed. EEG measurements were completed before, during, and after SH engaged in his "reading of another" which we called the "interpretational state" (IS) because of the unique pattern of EEG activity (HC). To discern if the HC was quantitatively associated with SH's rated accuracy, 10 photographs of related individuals (children, niece, grandparents etc.) were supplied by 4 different people (Male, N=4, mean age = 24.5 SD = 1.12) who were not present at the time of the experiment but whom he had met previously. During the interpretations his comments were recorded by audiotape and his QEEG was continuously measured. SH was given each photograph by one of two people sitting with him in an acoustic chamber (the other person monitored the portable EEG measurements).

When his comments were finished the photograph was removed and another one was handed to SH. He would look at the photograph of the person for about 2-3 s and then begin with a series of brief descriptors concerning health, education, friendship patterns, death in the family, history of diseases, and general "personality". The proportions of approximately 10 to 15 comments (each comment about 2 s) made for each of the 10 pictures from each person (n=40 photographs) were scored for specific accuracy by each of the four people separately a week later when the comments had been transcribed. The categories for the statements by SH were rated as false, generally true, specifically true, future references, and "don't know" or "unable to confirm". A relative score for each picture was computed by calculating the ratio of true (sum of generally and specifically true) to the total of false plus true. A second proportional score was compute by calculating the number of true statements divided by the total number of comments for each picture. These scores were correlated with the total duration and number of HCs per picture. Factor analyses (varimax rotation) were completed utilizing the two dependent measures. Multiple regression analyses were also completed for these two dependent variables as a function of the various EEG spectral power measures. As a control for movement and muscle artifacts we instructed SH to complete his reading or "call his angel" in the same manner that he would when interpreting someone's photograph.

2.2.3. Microstate and Accuracy Correlations

Research by Koenig et al (2002) has revealed consistent microstates whose durations are within the range of a percept, approximately 80 to 120 msec. The microstates are defined as periods of quasi-stable topographical maps of the cumulative scalp electrical field disregarding polarity (Lehmann et al. 2010). Isopotential lines reveal four major stable maps of microstates that accommodate almost three-quarters of the variance in voltage fluctuations (Figure 2a). These four patterns are remarkably consistent across ontogeny and vary primarily in their durations and proportions (Koenig et al. 2002) and may reflect a type of information patterning or processing (Lehmann et al. 2010).

The microstates were extracted from eye blink artifact free EEG records taken from the 19 channel EEG during the reading and analyzed according to Koenig et al's (2002) procedures. The characteristics of the microstates during SH's rest states and IS were mapped. The durations and numbers of specific states per unit time during his "readings" of the 40 photographs were correlated with the measures of accuracy noted in the previous section. Factor analyses (varimax rotation) were also completed between the different microstate measures and the proportional and relative accuracy score measures.

2.2.4. Temporal Discrimination ("Sense of Now")

The sense of now can be defined as the minimum amount of time required to discriminate between two discrete stimuli. Previous results from our lab have identified that this minimum time required is about 28 ms based on the following procedures (Dotta et al.,in submission). To discern if SH's temporal discrimination of "now" was different, he was instructed to identify which of two red circles appeared first on a laptop computer screen. The two circles each had a diameter of 1cm and separated by a distance of 5cm on the screen. The circles were randomly presented with respect to both place presentation (left first, right first) and temporal presentation (order of duration segments). SH completed this paradigm while he sat in a comfortable chair in a dimly lit acoustic chamber (the same chamber in which he interpreted the photographs of 2.2.2). The duration of delay between the two symbols ranged from 5 to 40 msec (5 blocks of 19 trials). The time at which he displayed 100% accuracy for discrimination between two circles was considered his threshold for "now".

2.2.5. EEG Responses of Subjects Proximal to Harribance

Journal of Consciousness Exploration & Research| December 2010 | Vol. 1 | Issue 9 | pp. 110-128
Hunter, M.D., Mulligan, B.P., Dotta, B. T., Saroka, K. S., Lavallee, C. F., Koren, S. A., & Persinger, M. A., *Cerebral Dynamics and Discrete Energy Changes in the Personal Physical Environment During Intuitive-Like States and Perceptions*

115

Historically, the proportion of accuracy of information about others has been considered greatest when SH was sitting within about 1 to 2 m of the person. We reasoned that if the source of the information originated from the space occupied by the participants, specific changes in their brain activity as inferred by QEEG may occur when SH was displaying the IS compared to baseline or other conditions. On separate days, the QEEGs of three different individuals (Male, N=3, mean age 24 SD 0.5) (not involved with part 2.2.2) were recorded while SH's EEG was being measured and he was interpreting their photographs. Each of the individuals and SH had 20, 2 second artifact free segments extracted for analysis of coherence. The subjects were assessed using the 8 channel EEG system while SH was assessed using the 19 channel system. Synchronization of EEG records were completed through 3 stages; 1) both computers were updated through Windows XP's synchronized internet time server every 1 hour. Any record taken was given a digital time stamp; 2) throughout the records stopwatches were concurrently synchronized to the same international time stamp and were used to further synchronize events within the record; 3) Each interpretation session had digital audio recording which was also time stamped and was used to transcribe the events and the associated timeline.

2.2.6. Photon Emission During the Interpretational State

To discern if changes in photon emission were occurring in the proximity of the scalp (inferred as a result of brain activity), SH sat in a comfortable chair (in a different room on two successive days) in the dark while photon emission was recorded from the right side of his head. We selected the right side on the same plane as his parietal and temporal lobes because: 1) previous research demonstrated enhanced SPECT activity in this region (Roll et al, 2002), and, 2) it was the locus of the highly consistent EEG configuration that was associated with the accuracy of the ratings for his interpretations.

A Model 15 Photometer from SRI Instruments (Pacific Photometric Instruments) with a PMT housing (BCA IP21) for a RCA electron tube (no filters) was calibrated by comparing directly to a digital luxmeter at higher intensities (> 1 lux) and by measuring the response to a 700 nm LED at 10 mA (5 millicandella; 2 millilumens/45 degree) at various distances for intensities of less than 1 lux. Lux was transformed to Watts/m^2. The output was transformed to mV (millivolt meter) and sent to an IBM ThinkPad laptop (Windows 95) where samples were taken 3 times/s during the experimental periods. Calibration indicated that a change of 1 unit at this sensitivity was equivalent to about 5×10^{-11} W/m^2 or a total energy of about 1.5×10^{-11} J/s at this distance.

2.2.7. Proximal Changes in the Geomagnetic Field Intensity

The changes in intensity (in nanoTesla) within the earth's magnetic field in the three spatial planes surrounding SH's head were measured. Routine measurements in our laboratory as well as others have revealed that the semiconductive properties of the human mass can alter the shape and intensity of the geomagnetic field along the boundaries of the body (Presman 1970). We suspected that such changes may occur when he was engaging quietly in the IS compared to rest conditions. SH completed his IS while maintaining silence to eliminate any resultant activity with respect to movement.

To measure possible left vs right hemispheric differences and how they might affect the immediately adjacent magnetic field, simultaneous measurements were taken over the left and right parietal region of the skull by placing the magnetometer sensors over each side. To discern changes in the earth's magnetic field proximal to the right side of SH's head, the two separate MEDA FVM-400 magnetometers were placed on narrow aluminum frames so the sensors would be located on the same plane as his right temporoparietal region at distances of 25 cm and 1 m. The magnetometers' outputs (X,Y,Z) were recorded by IBM laptops. Sampling (17.4 samples/s) durations were 30 sec by MEDA software through an IMB ThinkPad laptop. To compare the effects of movement artifacts, SH was instructed to move his right arm quickly and to point upwards. These conspicuous artifacts, which were qualitatively different from the distinct changes that occurred when he was sitting quietly and engaging in the IS, displayed the usual noisy oscillatory excursions with a maximum change between 100 and 200 nT.

2.3 Statistical Analyses

All analyses, including correlations (Pearson and Spearman), factor analyses, and multiple regression were completed by SPSS software.

3. Results

3.1. Quantitative Electroencephalography (QEEG)

The most unique characteristic of SH's brain activity was brief (a few seconds) high frequency (gamma range, i.e., 30 to 40 Hz) consistent configurations over the right rostral-parietal-temporal (C4,T4) and the right central and orbital frontal region (Figure 1a) while he was "calling the angel"; other areas showed slower (irregular 8 to 15 Hz) state-expected activity. When the pattern was corrected for talking (calling of the angel only) (Figure 1b), the configuration was specific to the right frontal (F4,F8), central (C4) and temporal (T4) locations, particularly around 20 Hz for F8 and T4. sLORETA software (Figure 1d, 1e) showed enhanced activity that extended into the insula and temporal cortices in the right hemisphere when compared to his resting baseline ($p < 0.001$). This pattern defined the Harribance Configuration (HC).

The HC occurred reliably for a total of about 20 s per min and with an average of 6 separate episodes when he was interpreting people or photographs but not when he engaged in other activity such as resting, meditating, or spontaneous talking about ongoing events. During interpretations SH reported he sensed a "sentient presence" to which he attributed his "information" about the participant or associated photographs. He reported he could control this "presence", even when he was not talking, which has been subjectively apparent since early childhood. Fourier analyses of the power output (uV^2/Hz) of his brain during his interpretations were associated with increases in all lobes within the delta (1 to 4 Hz) and theta (4 to 8 Hz) range upon which the HC was superimposed (Figure 1f). The increases in power within the delta and theta band were most apparent over his right temporal lobe.

Journal of Consciousness Exploration & Research| December 2010 | Vol. 1 | Issue 9 | pp. 110-128

117

Hunter, M.D., Mulligan, B.P., Dotta, B. T., Saroka, K. S., Lavallee, C. F., Koren, S. A., & Persinger, M. A., *Cerebral Dynamics and Discrete Energy Changes in the Personal Physical Environment During Intuitive-Like States and Perceptions*

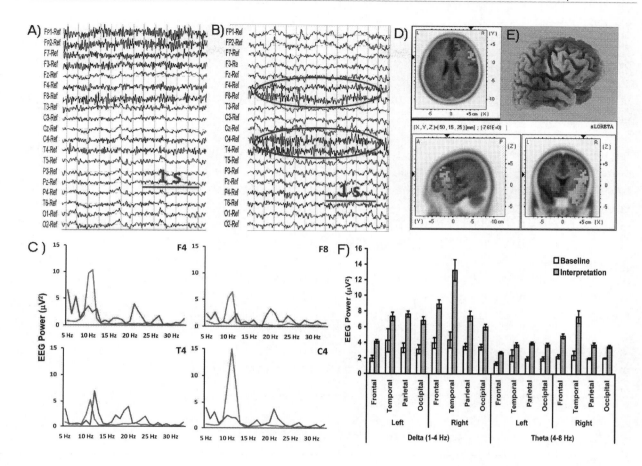

Figure 1. Samples of electroencephalographic patterns and software representations of the "Harribance configuration" (HC) during his interpretations. (a) Electronic monopolar recordings from 19 channel EEG while SH was interpreting a subjects photograph (b) Electronic monopolar recordings from 19 channel EEG while SH was in the IS but not talking (HC indicated by elliptical regions). (c) Power spectra (blue while talking, red without talking during the "interpretational state", IS) showing major enhancements within the right frontocentral and temporal lobes at approximately 20 Hz. (d) s-LORETA profile showing specific activation within the gamma band (light blue) during the HC within primarily the right temporal-insular region. (e) s-LORETA profile of major activation within the gamma band (blue) during the IS (while talking). (f) General increase in power within the delta and theta ranges during the HC compared to baseline for major areas over both hemispheres. Note the particular enhancement over the right temporal lobe.

3.2. Rated Accuracy of Types of Interpretations and EEG Patterns

Raters ranked all comments for each picture individually during the approximately 1 min interpretation (total comments M=18, SD=4.7) according to five categories. The categories and proportions (means with standard deviations in parenthesis) for the 40 photographs were: false

Journal of Consciousness Exploration & Research| December 2010 | Vol. 1 | Issue 9 | pp. 110-128
Hunter, M.D., Mulligan, B.P., Dotta, B. T., Saroka, K. S., Lavallee, C. F., Koren, S. A., & Persinger, M. A., *Cerebral Dynamics and Discrete Energy Changes in the Personal Physical Environment During Intuitive-Like States and Perceptions*

118

7(10) %, generally true 42(19) %, specifically true 25(18) %, (generally true and specifically true were summed as true 67(20) %), don't know 9(9) % and future reference 18(15) %. The correlation between the relative accuracy for his comments for the photographs and the numbers of HCs was r = 0.46 (p <0.001). The magnitude of the association between the numbers of HCs and the proportion of true statements was similar. The raters were blind with respect to the measures of the HC for each photograph.

Factor analysis (loading coefficients in parentheses) of the total number of false (-0.02), true (0.80), don't know (-0.52), future (-0.58) and relative numbers of HCs (0.74) indicated that only HCs and true ratings shared the same positive variance and explained 37% (Eigenvalue = 1.89) of the variance. Factor analyses (all after varimax rotation) showed similar loadings when the relative duration of HCs (total duration of HCs/total time of trial) was substituted for the relative number of HCs and for relative true (0.78), relative false (-0.15), total statements (0.11), relative number HCs (0.76) and relative duration of HCs (0.71) were included. Multiple regression analyses with each of the 1 Hz increments of power showed that both the duration of the HC (Table 1) and rated accuracy (Table 2) could be predicted by specific bands over the right frontal, central, and rostral temporal regions.

Table 1. Step-wise multiple regression of the duration of HCs (dependent variable) during 40, 1 min samples while SH was interpreting for all frequency bands (1 Hz to 40 Hz) within F4, F8, C4, and T4

Variable	R	R^2	ΔR^2	B	Beta
F4 27 Hz	0.76	0.58	0.58	9.27	0.62
C4 16 Hz	0.85	0.72	0.16	-19.28	-0.30
F8 21 Hz	0.87	0.76	0.04	- 4.59	-0.32
T4 12 Hz	0.89	0.80	0.04	11.03	0.18
F4 26 Hz	0.91	0.83	0.03	4.96	0.30
Constant				14.55	

Table 2. Step-wise multiple regression of the relative accuracy scores (dependent variable) during 40, 1 min samples while SH was interpreting for all frequency bands (1 Hz to 40 Hz) within F4, F8, C4, and T4

Variable	R	R^2	ΔR^2	B	Beta
T4 17 Hz	0.32	0.11	0.11	0.20	0.40
C4 33 Hz	0.60	0.36	0.25	-1.15	-0.70
T4 19 Hz	0.66	0.44	0.11	0.16	0.40
Constant				0.86	

3.3. Microstates and Accuracy Correlations

As shown in Figure 2a, during resting (eyes closed) the structure of SH's fundamental microstates did not differ appreciably from the normative data (Koenig et al, 2002); however, the duration of each of his states was conspicuously (about half) shorter (occurrence equivalent to about 7 Hz, rather than 3 to 4 Hz). During the IS or while he was interpreting photographs different states emerged that have not been reported in the literature (Figure 2b). Class A (red) and to some extent

Journal of Consciousness Exploration & Research| December 2010 | Vol. 1 | Issue 9 | pp. 110-128   119
Hunter, M.D., Mulligan, B.P., Dotta, B. T., Saroka, K. S., Lavallee, C. F., Koren, S. A., & Persinger, M. A., *Cerebral Dynamics and Discrete Energy Changes in the Personal Physical Environment During Intuitive-Like States and Perceptions*

Class D (yellow) showed the same polarity in both cerebral hemispheres while Class B (green) and Class (C) (blue) showed the more typical but restricted bipolarity but within the same (right) hemisphere. The occurrence (expressed as Hz) for these states were consistent over the quartiles of the 60 sec interpretations. During his viewing of the 40 pictures the proportions of time covered by each of the four anomalous microstates were: 25(13) %, 32(11) %, 18(17) % and 25(7) %, respectively.

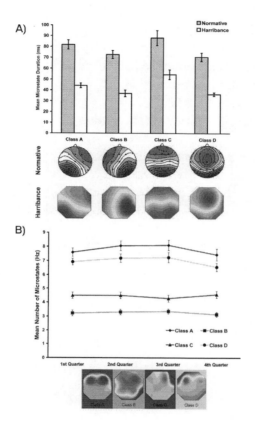

Figure 2. (a) Mean durations and SEMs (vertical bars) for each of the 4 classes of cerebral microstates for the normative sample and for SH. The topographic maps (looking down from the top with the person facing the top of the page) showed microstates when he was not engaging in interpretation that were similar to normative data. Red and blue indicate areas of opposite polarity. Normative maps adapted from reference (Koenig et al. 2002). (b) The frequency (inverse of duration) of each of the four anomalous cerebral microstates that emerged during the IS. Means and SEMs, based on 40 samples, are given for each state and are indicated by lines of the same color as the rectangle surrounding the maps. The four classes of microstates explained 54.58% (SD 10.46) of the variance in the analyzed epochs.

Factor analyses of the 40 scores for the pictures with the proportion of time for each state, the occurrence of the HC, and relative accuracy, revealed a factor (Eigenvalue = 2.16) loaded by class C (opposite polarity over right prefrontal and right temporoparietal lobes: 0.67), Class A (-0.79), the HC duration (0.82) and the relative accuracy rating (0.62). The most unusual microstate, the one

Journal of Consciousness Exploration & Research| December 2010 | Vol. 1 | Issue 9 | pp. 110-128

Hunter, M.D., Mulligan, B.P., Dotta, B. T., Saroka, K. S., Lavallee, C. F., Koren, S. A., & Persinger, M. A., *Cerebral Dynamics and Discrete Energy Changes in the Personal Physical Environment During Intuitive-Like States and Perceptions*

120

associated with the same polarity in both hemispheres (Class A), was correlated (0.55) with the numbers of statements that could not be verified by the raters. The mean duration for SH's normal microstates (when he was not engaged in the IS) and the states positively associated with the raters' accuracies was about one-half that of the normal person.

3.4. Temporal Discrimination (Sense of "Now")

SH's threshold for sense of now was not within the normal range. Whereas accuracies for normative samples of people were 100% when the discrepancies were above 28 msec, SH's most accurate scores occurred for time delays between 10 to 20 msec; above this interval (to a maximum of 40 msec) SH's accuracy was significantly less.

3.5. EEG Responses of Subjects Proximal to Harribance

All individuals displayed increases in beta and gamma activity (Figure 3a) over their temporal lobes in the presence of SH during the interpretations. The relative increase in power within this range compared to the baseline (when SH and the subject simultaneously engaged in eyes-closed relaxation) is shown in Figure 3b. In addition to the general increase in power within the right parietal and left temporal lobes of the participants, there was a conspicuous decrease in power within the 1 to 9 Hz range (with the exception of the 6 Hz to 7 Hz increment) over the right parietotemporal regions compared to these lobes in the left hemisphere.

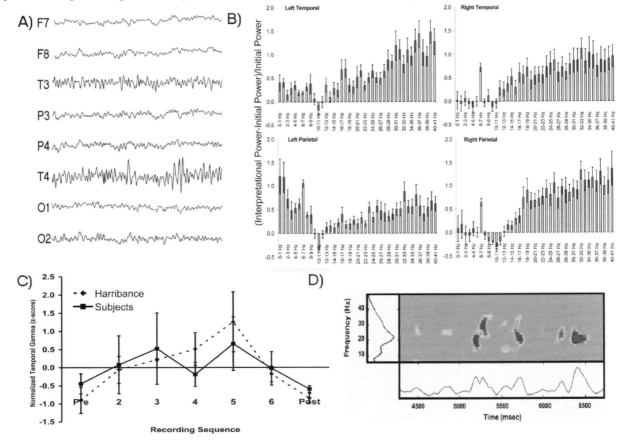

Figure 3. Responses in the brain activity of participants sitting (about 1 to 2 m) near SH while he was interpreting their photographs. (a) Typical EEG patterns of a participant while SH was interpreting their photographs. Note the higher frequency and amplitude of activity over the left (T3) and particularly the right (T4) temporal lobes. (b) Average relative change in power within each 1 Hz increment of brain activity between 1 Hz and 41 Hz for four participants during SH's interpretation relative to baseline. Note the marked decrease in power (< 10 Hz) within the right temporoparietal region except for the 6-7 Hz band within the participants' records when SH was engaged in the IS. (c) Averaged power outputs for SH and participants within the gamma range (temporal lobe) during SH's pre-IS, IS, and post-IS for the 40 photographs. Vertical bars indicate SEMs. (d) Cross-spectral analyses of power (yellow=mild coherence; red= maximum coherence) within the left temporal lobe of a sample subject and SH's right temporal lobe within various frequency intervals over time (milliseconds).

There was increased similarity of the EEG patterns for SH and the subjects over the temporal lobes (average of left and right) in particular. During the pre-IS, IS, and post-IS periods the relative power of the two brains did not differ significantly within the gamma range (Figure 3c); this was not evident for delta, theta, low-alpha, high alpha or beta bands. The increased power within the 33 to 35 Hz range over the temporal lobes of SH and the participants was conspicuous even though the latter talked rarely. Cross spectral analyses of the raw data (20, 2 sec samples per participant) indicated increased coherence within a narrower 19 to 20 Hz band as well as the wider 30 to 40 Hz band within SH's right temporal lobe and the participants' left temporal lobes during his interpretations (Figure 3d). A similar coherence occurred between SH's left and right temporal lobes. The means and standard deviations for the duration of the coherence were $M = 137$ msec and $SD = 55$ msec (or about 7 Hz) while these values for the inter-event intervals were 587 (475) msec.

3.6. Photon Emission During the Interpretational State

As can be seen in Figure 4a, the energy density (Watts/m^2) as measured by a photomultiplier tube at 0.15 m along the right side of SH's head while he sat in a comfortable chair in complete darkness, showed an increase of about 1 unit of average transmittance during the IS compared to resting baselines. The changes were reversible (Figure 4b), replicable on two separate days (Figure 4c) and controllable by the experimenters' instructions for SH to start or stop the IS. The latencies between the instruction for SH to begin the IS and the increases in transmittance were between 10 and 20 sec. Spectral analyses indicated that microvariations associated with these shifts displayed primary power peaks according to Fourier analyses between 0.2 and 0.4 Hz. Calibration indicated that a change of 1 unit at this sensitivity was equivalent to about 5×10^{-11} W/m^2 or a total energy of about 1.5×10^{-11} J/s at this distance.

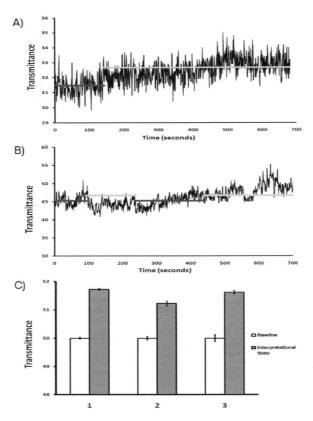

Figure 4. Photomultiplier tube measurements demonstrating an increase of approximately 1 unit of photon emission ("transmittance") during IS. (a) Example of the increase of approximately 1 transmittance unit (yellow vs red) about 10^{-11} J/s) at 0.15 m from SH's right hemisphere when engaged in the IS (yellow line) after baseline (red). (b) Example of the reversibility of power density when SH stopped (red line) and initiated (yellow line) the IS. (c) Means and SEMs for the changes in photon detection between baseline (resting) and after initiation of the IS for three separate experiments.

3.7. Proximal Changes in the Geomagnetic Field Intensity

The steady-state (static) geomagnetic field in the horizontal plane was about 2,000 nT less over SH's right temporoparietal region compared to his left. This was similar to what was measured about ten years ago when he visited the laboratory. During the IS (Figure 5a-c) there were relatively sudden changes in the intensity of the local geomagnetic field within 1 cm of skull. They began about 10 to 20 sec after the instructions to begin the IS. The typical magnitude of these changes was about 150 nT. According to the classic equation: $J=(B^2/2\pi u)$ multiplied by the volume (of his cerebrum), the energy available in this change in the geomagnetic field during the IS would be about 10^{-11} J.

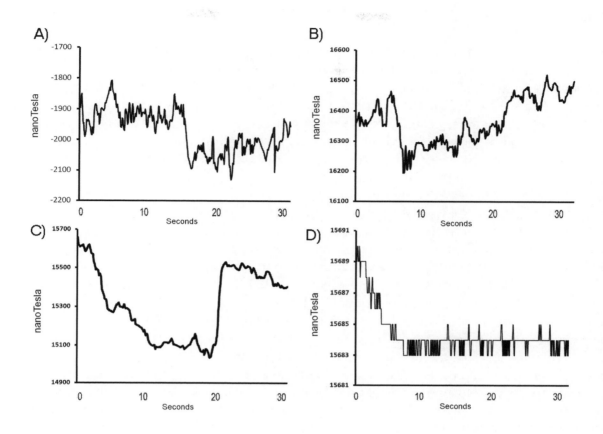

Figure 5. Changes (in nanoTesla) in the intensity of the earth's magnetic field (showing sample X,Y,Z components, measured in nanoTesla) around SH for his right temporoparietal region (a-c, at 1 cm; D=1 m). (a) Change (in nT) in geomagnetic intensity (horizontal plane in direction of magnetometer) 1 cm from the right side of SH's head about 15 s after instruction for IS. (b) Change in nT beginning about 15 s after instruction showing decrease and recovery. (c) Changes in intensity beginning at the onset of the IS effect and recovery when IS stops after about 20 s. (d) Decrease in intensity (horizontal plane) at 1 m during onset of IS. The spiky transients reflect the resolution (1 nT) of the instruments.

In the second set of experiments when the IS was engaged and the geomagnetic field was measured at 0.25 and 1 m from the right side of his head there were clear decreases of about 15 nT and 5 nT (Figure 5d), respectively, during the state compared to when he was instructed to relax. The energy change, according to the expanding volume determined by the distance, would remain equivalent to about 10^{-11} Joules. Fourier analyses indicated intrinsic amplitude modulations during these shifts that peaked within the 0.2 Hz to 0.6 Hz range.

4. Discussion

SH attributes his interpretations and intuitions to information from a sensed presence. Variants of a sensed presence have included the ancient Greek's Muses (Persinger and Makarec, 1992), mystical or religious "spirit guides" (Evans 1984) and even the concept of the "hidden observer" following hypnosis (Harrington 1995). The latter has been hypothesized to allow experients to solve complex problems even though they cannot recall the source of the solutions. These experiences have strong associations with right hemispheric processes, creativity and intuition (Chavez-Eakle et al. 2007; Jung Beeman et al. 2004). It is important to emphasize that all of these "complex" behaviors are derived from specific patterns of brain activity.

Our results indicate that reliable, measurable changes in the brain activity within the right parietotemporal region of SH occurred when he was interpreting another person or a photograph. These changes were consistent while engaging in the IS and without a talking artifact present. This is the same area that had displayed marked uptake in tracer during SPECT (Roll et al. 2002) which is not susceptible to the muscle artifacts. Given the sLORETA profile of enhanced activity within the right medial temporal, parietotemporal and insula regions (Figure 1d), we suggest that SH's ability is a variant of the process that contributes to normal intuition. It has been associated with the acquisition of information regarding past, present and future events that involve complex inferencing about which the experient is allegedly unaware. We suggest that individuals like SH are at the opposite end of the spectrum compared to Asperger's Syndrome (Senju et al, 2009) for the capacity to engage in Theory of Mind. Whereas the former display "mind-blindness", Mr. Harribance may display a "hypermentality" that may allow sensitivity to information and environmental cues not typically accessible by the normal person.

The shared variance between the ratings of the relative accuracy of the 40 photographs and the numbers of HCs during the interpretation of each photograph further validates the relationship between the EEG and SH's experience. The convergence suggests a third factor may be associated with SH's brain activity and the accuracy of information stated for each photograph. To understand what alteration in his consciousness might be occurring, the HC was analyzed for microstates of integrated cerebral activity (Koenig et al. 2002). If we assume that the neuroelectromagnetic correlates of consciousness are related systematically to the manner in which the billions of potential events (Norretranders 1998) penetrating the brain are detected (become stimuli) and are integrated into experiences, alterations in these functions might allow access to information not available in normal states.

The alteration in what is likely to be SH's sense of "now" suggests a different temporal processing or integration of sensory stimuli not typically seen in the average person. SH could discriminate latencies in the differential serial onset of two stimuli at durations between 10 and 20 msec where the average person requires more than 28 msec with this paradigm. This "doubled" processing rate was also reflected in the net durations of his microstates. The coherence was equivalent to about 7 Hz which was similar to the value for coherence between SH's left and right temporal lobes.

In addition to the normal microstates, SH also displayed patterns that have not been previously reported. The "double (same) polarity" pattern (Class A) between the two hemispheres might be interpreted as a brief period of marked intercalation or "integrated state". Our hypothesis (Booth et al, 2005; Persinger, 1993) has been that such transient interactions allow conscious access to primarily right hemispheric information that is then transformed to linguistic equivalents or images (Persinger et al, 2002). We will be exploring the possibility that the correlations observed in this study between the non-rateable comments and the "double polarity" pattern may have contained information that was not accessible to the rater but may be accurate according to the first-person experience of the person within the photographs.

That there were congruent changes in power within the 33 to 35 Hz range over the temporal lobes of SH and the participants in the presence of SH offers an alternative explanation to the long history of belief that individuals with SH's capacity are accessing the cognitive processes of distant or deceased individuals. The experiences and memories of the participant or experient is the primary reference by which the statements about others are verified. One interpretation of the increased coherence between the 19 to 20 Hz band and 30 to 40 Hz band over SH's right temporal lobe and the participants' left temporal lobes during his interpretations would be an as yet unspecified direct access to the information within the participants' cerebral space. A more parsimonious explanation would involve a variant of enhanced suggestibility by his proximity.

If the former explanation is valid, then SH's discernment of this information would involve physical mechanisms. We examined this possibility in two ways. We measured the quantitative increase in photon emission from the right side but not the left side of his brain during his IS. University students tested in our laboratory also show increased photon emission over the right hemisphere but not over the left when they imagine light. SH's values were 4 times higher (about 6 standard deviations above the mean). We also measured the changes in the static component of the geomagnetic field around the right side of his skull. The approximately 2,000 nT reduction in the static intensity of the geomagnetic field along the right side of his skull at the level of the temporoparietal lobe compared to the left would be consistent with greater penetration of flux lines within the skull and presumably his cerebral space.

We suggest such penetration into SH's right temporoparietal region would allow cerebral access to potential information that might be contained in a yet to be specified format within the geomagnetic field (Persinger, 2008). The increment of energy associated with this access would be in the order of 10^{-11} J. This quantum of energy was emitted from the right side of his brain during the IS. In addition 10^{-11} J was the equivalent energy associated with the changes in geomagnetic intensity within the volume of space 25 cm and 100 cm from his skull. The increases in photon emission and decreases in geomagnetic intensity were matched for magnitude of energy, latency to be displayed and superimposed amplitude variation (between 0.2 Hz to 0.6 Hz). The convergence of values does not prove information was transferred from the environment but does support the first law of thermodynamics (conservation of energy), that energy is neither created nor destroyed but changes form, which may be relevant to the Harribance phenomena.

The value of approximately 10^{-11} Watts (J/s) is also well within the range generated by the electromagnetic component of brain activity. The influence of a single action potential's net change of 1.2×10^{-1} V (120 mV) upon a single charge of 1.6×10^{-19} A s is about 2×10^{-20} J (Persinger et al,

Journal of Consciousness Exploration & Research| December 2010 | Vol. 1 | Issue 9 | pp. 110-128   126

Hunter, M.D., Mulligan, B.P., Dotta, B. T., Saroka, K. S., Lavallee, C. F., Koren, S. A., & Persinger, M. A., *Cerebral Dynamics and Discrete Energy Changes in the Personal Physical Environment During Intuitive-Like States and Perceptions*

2008). The value of 10^{-11} J would be equivalent to about 10^9 action potentials or 1 billion neurons displaying one action potential per sec. Assuming an average frequency of about 10 Hz per neuron, this would involve in the order of 100 million cortical neurons. This value is well within the range estimated to be activated within the temporoparietal cortices.

5. Conclusion

We suggest that the unique organization of Sean Harribance's brain has allowed apparent access to information from others' memories. His accuracy has been sufficient to maintain his employment and be accessed by multiple private and government agencies. Quantitative EEG analyses indicated a fixed pattern of reliable increases in power over portions of the right hemisphere. The total numbers and durations of this configuration were significantly correlated with rated accuracies of information of people within photographs. SH displayed microstates that included normal as well as unique patterns whose durations were about half the values obtained for the average person.

The actual stimuli that became the information experienced by SH during his interpretations are not clear. The marked coherence of cerebral activity between SH and the participants during his close proximity and display of the IS indicates that a component of the information may originate from within the brains of the participants or a third factor shared by both. The discrete changes in photons concomitant with alterations in the intensity in the surrounding geomagnetic field in the vicinity of SH and the participants indicate involvement of physical mechanisms worthy of thorough exploration. In conclusion there is evidence that Sean Harribance intuits verifiable information about the history and status of others and that the processes are: 1) associated with discrete patterns of his brain activity, and, 2) consistent with the current understanding of the "Theory of Mind".

Acknowledgements: The authors wish to thank Dr. Ghislaine Lafreniere for proof reading and editing the manuscript. The authors would also like to acknowledge the Harribance Foundation for their contributions allowing Mr. Harribance to visit our laboratory.

References

Allison, T., Puce, A., and McCarthy, G. (2000). Social perception from visual cues: role of the STS region. Trends Cogn. Sci. *4*, 267-278.

Babayev, E. S. and Allahverdiyeva, A. A. (2007). Effects of geomagnetic activity variations on the physiological and psychological state of functionally healthy humans: some results of the Azerbaijani studies. Advances in Space Research, *40*, 1941-1951.

Brunet, E., Sarfati, Y., Hardy-Bayle, M.C., and Decety, J. (2000). A PET investigation of the attribution of intentions with a nonverbal task. Neuroimage *11*, 157-166.

Chavez-Eakle, R.A., Graff-Guerrero, A., Garcia-Reyna, J.C., Vaugier, V., and Cruz-Fuentes, C. (2007). Cerebral blood flow associated with creative performance: a comparative study. Neuroimage *38*, 519-528.

Dotta B., and Persinger MA. (In submission) Temporal discrimination between discrete stimuli represents our perception of "now".

Journal of Consciousness Exploration & Research| December 2010 | Vol. 1 | Issue 9 | pp. 110-128
127
Hunter, M.D., Mulligan, B.P., Dotta, B. T., Saroka, K. S., Lavallee, C. F., Koren, S. A., & Persinger, M. A., *Cerebral Dynamics and Discrete Energy Changes in the Personal Physical Environment During Intuitive-Like States and Perceptions*

Delorme, A., and Makeig, S. (2004). EEGLAB: an open source toolbox for analysis of single-trial EEG dynamics including independent component analysis. Journal of neuroscience methods *134,* 9.

Evans, H., (1984). Visions, apparitions, alien visitors: a comparative study of the entity enigma (Wellingsborough: Aquarian Press).

Fletcher, P.C., Happe, F., Frith, U., Baker, S.C., Dolan, R.J., Frackowiak, R.S., and Frith, C.D. (1995). Other minds in the brain: a functional imaging study of "theory of mind" in story comprehension. Cognition *57,* 109-128.

Fuchs, M., Kastner, J., Wagner, M., Hawes, S., and Ebersole, J.S. (2002). A standardized boundary element method volume conductor model. Clin. Neurophysiol. *113,* 702-712.

Happe, F. (2003). Theory of mind and the self. Ann. N. Y. Acad. Sci. *1001,* 134-144.

Harribance, C.C., (1994). Sean Harribance (Port of Spain: Sean Harribance Institute).

Harrington, A., (1995). In: Brain Asymmetry R. J. Davidson, K. Hugdahl eds. (Cambridge: MIT Press)

Hoffman, E.A., and Haxby, J.V. (2000). Distinct representations of eye gaze and identity in the distributed human neural system for face perception. Nat. Neurosci. *3,* 80-84.

Jung-Beeman, M., Bowden, E.M., Haberman, J., Frymiare, J.L., Arambel-Liu, S., Greenblatt, R., Reber, P.J., and Kounios, J. (2004). Neural activity when people solve verbal problems with insight. PLoS Biol. *2,* E97.

Jurcak, V., Tsuzuki, D., and Dan, I. (2007). 10/20, 10/10, and 10/5 Systems Revisited: their Validity as Relative Head-Surface-Based Positioning Systems. Neuroimage *34,* 1600-1611.

Koenig, T., Prichep, L., Lehmann, D., Sosa, P.V., Braeker, E., Kleinlogel, H., Isenhart, R., and John, E.R. (2002). Millisecond by millisecond, year by year: normative EEG microstates and developmental stages. Neuroimage *16,* 41-48.

Kuo, W.J., Sjostrom, T., Chen, Y.P., Wang, Y.H., and Huang, C.Y. (2009). Intuition and deliberation: two systems for strategizing in the brain. Science *324,* 519-522.

Lehmann D., Pascal-Marqui R.D, Strik, W.K., and Koenig T. (2010) Core networks for visual-concrete and abstract thought content: A brain electric microstate analysis NeuroImage *49,* 1073-1079

Liang-Cheng, T., Luo, J. and Gillies, G. T. (2005). The mass of the photon. Rep. Prog. Phys. *68,* 77-130.

Michon, A. L. and Persinger, M. A. (1997). Experimental simulation of the effects of increased geomagnetic activity upon nocturnal seizures in epileptic rats. Neurosci. Lett. *224,* 53-56.

Mulert, C., Jager, L., Schmitt, R., Bussfeld, P., Pogarell, O., Moller, H.J., Juckel, G., Hegerl, U. (2004). Integration of fMRI and simultaneous EEG: towards a comprehensive understanding of localization and time-course of brain activity in target detection. NeuroImage, *22*(1), 83-94.

Mulligan, B. P., Hunter, M. D. and Persinger, M. A. (2010). Effects of geomagnetic activity and atmospheric power variations on measures of brain activity: replication of the Azerbaijani studies. Advances in Space Research, *45,* 940-948.

Norretranders, T., (1998). The User Illusion (New York: Penguin).

Pascual-Marqui, R.D. (2002). Standardized low-resolution brain electromagnetic tomography (sLORETA): technical details. Methods Find. Exp. Clin. Pharmacol. *24 Suppl D,* 5-12.

Persinger, M. A. (1980). The weather matrix and human behavior. N.Y.: Prager.

Persinger, M. A. (2008). On the possible representation of electromagnetic equivalents of all human memory within the earth's magnetic field: implications for theoretical biology. Theoretical Biology Insights, 1, 3-11.

Persinger, M. A. (2010). 10^{-20} Joules as a neuromolecular quantum in medicinal chemistry: an alternative approach to myriad molecular pathways. Current Medicinal Chemistry (in press).

Persinger, M. A., Hoang, V. and Baker-Price, L. (2009). Entrainment of stage 2 sleep spindles by weak transcerebral magnetic stimulation of an "epileptic" woman. Electromagnetic Biology and Medicine, *28,* 374-482.

Journal of Consciousness Exploration & Research| December 2010 | Vol. 1 | Issue 9 | pp. 110-128

128

Hunter, M.D., Mulligan, B.P., Dotta, B. T., Saroka, K. S., Lavallee, C. F., Koren, S. A., & Persinger, M. A., *Cerebral Dynamics and Discrete Energy Changes in the Personal Physical Environment During Intuitive-Like States and Perceptions*

Persinger, M. A., Koren, S. A. and Lafreniere, G. F. (2008). A neuroquantologic approach to how human thought might affect the universe. NeuroQuantology, 6, 262-272.

Persinger, M.A., Meli, S., and Koren, S.A. (2008). Quantitative discrepancy in cerebral hemispheric temperature associated with "two consciousnesses" is predicted by neuroquantum relations. Neuroquantology *6*, 369.

Persinger, M.A., and Makarec, K. (1992). The feeling of a presence and verbal meaningfulness in context of temporal lobe function: factor analytic verification of the muses? Brain Cogn. *20*, 217-226.

Persinger, M. A. and Tiller, S. G. (2008). A prototypical spontaneous "sensed presence" of a Sentient Being and concomitant electroencephalographic activity in the clinical laboratory. Neurocase, *14*, 425-430.

Presman, A. S. (1970) Electromagnetic fields and life. Plenum: N.Y.

Popp, F. A. (1979) Photon storage in biological systems. In F. A. Popp, F. A., Becker, G.,, Konig, H. L., Pescha W. (eds) Electromagnetic bioinformation. Munich: Urban and Schwartzenberg, pp.123-149.

Roll, W.G., Persinger, M.A., Webster, D.L., Tiller, S.G., and Cook, C.M. (2002). Neurobehavioral and neurometabolic (SPECT) correlates of paranormal information: involvement of the right hemisphere and its sensitivity to weak complex magnetic fields. Int. J. Neurosci. *112*, 197-224.

St-Pierre, L. S. and Persinger, M. A. (2006). Experimental facilitation of the sensed presence is predicted by specific applied patterns of applied magnetic fields and not by suggestibility: re-analysis of 19 experiments. Int. J. Neurosci. *116*, 1-18.

Saxe, R., and Kanwisher, N. (2003). People thinking about thinking people. The role of the temporo-parietal junction in "theory of mind". Neuroimage *19*, 1835-1842.

Saxe, R., and Wexler, A. (2005). Making sense of another mind: the role of the right temporo-parietal junction. Neuropsychologia *43*, 1391-1399.

Senju, A., Southgate, V., White, S., and Frith, U. (2009). Mindblind eyes: an absence of spontaneous theory of mind in Asperger syndrome. Science *325*, 883-885.

Sun, Y., Wang, C., and Dai, J. (2010). Biophotons as neural communication signals demonstrated by in situ biophoton autography. Photochemical and Photobiological Sciences, DOI: 10.1039/b9pp00125e.

Vogeley, K., Bussfeld, P., Newen, A., Herrmann, S., Happe, F., Falkai, P., Maier, W., Shah, N.J., Fink, G.R., and Zilles, K. (2001). Mind reading: neural mechanisms of theory of mind and self-perspective. Neuroimage *14*, 170-181.

Winterer, G., Mulert, C., Mientus, S., Gallinat, J., Schlattman, P.,Dorn, H., Herrmann, W.M. (2001). P300 and LORETA: comparison of normal subjects and schizophrenic patients. Brain Topograpny, *13*(4), 299-313.

Essay

Cutting through the Enigma of Consciousness

Chris King[*]

ABSTRACT

Critical to the investigation of consciousness is that it is existentially completely different from the objective physical world description, being experienced directly only by the subject, and not being subject to the same criteria of replicability a physical world experiment has. Also the observer cannot control their consciousness objectively in the same manner a physical experimentalist can their equipment, because any attempt to change consciousness carries the observer into a new conscious situation as well. In this respect the exploration of consciousness has similarities to quantum measurement. This renders all forms of introspection made as if we are looking at consciousness objectively, completely, or partially invalid.

Key Words: consciousness, enigma, observer, subjective, physical world, quantum measurement.

The inner space of consciousness is sometimes able to perceive kaleidoscopic
'mindscapes', as if they are genuine perceptions of a 'world out there'.
How does the brain evoke these realities and what is their status,
By comparison with the subjective experiences we have of the real world?

This article is an exploration of where discovery about the human understanding of consciousness might be headed and why looking for answers may require a completely novel approach to understanding reality, different from anything we have encountered so far.

[*] Correspondence: Chris King http://www.dhushara.com E-Mail: chris@sexualparadox.org

The Scientific Lesson for Subjective Consciousness

In scientific terms, subjective consciousness remains the one phenomenon for which the description of physical reality has at this point not even the beginning of an explanation for. Although we know our subjective experiences are somehow a product of our brain states, we really have no idea of how a bunch of neurons firing off electrical impulses can come to generate all our conscious perceptions, dreams, memories and reflections of the world around us with all their diverse attributes, each of which is as indescribably different as a kaleidoscopic pattern of colour or a living landscape is to a musical symphony or even a complex cacophony of natural sound.

However the lesson of the scientific revolution shows us some important potential features of the quest for understanding consciousness that may be key to making real progress. The scientific revolution didn't come easily, because nature revealed itself to work in subtle ways that violated the simplistic assumptions of traditional, and particularly religious thought. It turned out that the Earth was neither flat, nor the centre of the universe, which, rather than being an airy heaven, in which angels with feathery wings could dwell, has turned out to be a maelstrom of black-holes and galactic collisions, in which life can take a foothold only on the surfaces of small rocky planets around small sun-like stars.

Even more perturbing, all life, far from being created by God, like clockwork toys in his image, appears to have emerged spontaneously from the slime, in a de-novo chemical synthesis, followed by the hit-and-miss process of mutational evolution, with humanity gracing the planet in a tortuous sexually-procreative journey of successful mutation through fish, reptiles and monkeys, a scenario which remains to this day the bane and nemesis of religious fundamentalists.

To cap the bag, we now understand the universe to be created almost from nothing, in a symmetry-broken cosmic inflation, whose mathematical complexity defies our imagination and ingenuity, despite many valiant ongoing efforts. As well, fundamental physics has entered into the mysterious territories of quantum uncertainty and quantum entanglement, altering forever our classical notions of temporal causality and physical reality.

We need to learn from the lessons of science's attempts to discover the nature of the real world, which has challenged our best minds through the centuries, and open ourselves to the possibility that consciousness, as we know it, is at least as unfamiliar to our preconceived notions as the physical universe has proved to be.

The Existential Dilemma and its Traditional Approaches

Nevertheless, the nature of consciousness is an urgent question which plunges right into the crucible of our psyche, because it leads to the ultimate existential anxieties: "What happens to me when my physical body dies?" "Is there any meaning in life if there is no after life after death?" "Is there a God looking after our fate?" "Am I a tiny part of a cosmic mind?" "Is there anything out there that cares, or are we just ships passing in the night alone, despite our delusions of love and togetherness amid frank exploitation of one another and of the natural world?"

For all the apparent solidity of the physical and biological world description, we are and remain throughout our loves conscious sentient beings, and it is only through the conduit of subjective consciousness that we come to witness the physical universe at all. And it is the stream of our subjective conscious impressions of reality that are all we have and that in which all our dreams and hopes and fears are enmeshed, despite the world we consciously perceive around us.

Dreaming can evoke bizarre realities which, despite seeming to be physically impossible, are palpably real (Oscar Dominguez "Memory of the Future").

This brings us rapidly back to the traditional answers to the existential dilemma, which present themselves most dominantly as religious beliefs. The monotheistic myth goes roughly as follows: "Yes there is a God - in fact the one true God of reality acting in history, unlike those other pagan idols, and who, despite being the creator of the entire universe, is also a moral deity who is 'jealous' of our fidelity to Him and might cast us into hell fire if we stray from unswerving belief in His power, majesty and commandments.

Despite the protestations of religious believers, this model of conscious existence is fatally flawed, because we now know that morality is a social manifestation, which takes root in a species as an evolutionary strategy which enhances inter-group dominance by reducing intra-social strife. In no way can any culturally-derived or revealed doctrine of moral causality be dominant over the reality of evolution and the wide variety of ecological niches evolution fills, from nutrient-giving plants through herbivores and carnivores to parasites and diseases.

The idea of a God which created nature and the entire universe stipulating any sort of moral imperative, let alone a final eschatology, is in complete contradiction to the open-ended indiscriminate mutational exploration and sheer creativity of the evolutionary paradigm, just as is the idea of a God creating the entire physical universe being in any way jealous of our fidelity is a contradiction in terms.

While the notion of a moral deity is a false projection of human cultural, sexual and social imperatives, the notion of a fall from paradise culminating in an apocalyptic awakening, amid heroic redemption, is a valid part of our collective emergence. The fall and awakening are to a considerable extent a real time description of our collective falling out, across the generations, from gatherer-hunter interdependence with nature, through the rise and fall of civilizations amid tumult and discord, to the present explosion of scientific knowledge (and technological and commercial exploitation of the planet) - a process of continuing culture-shock, arriving on the horizon at a greater understanding of our place in the universe.

Religious visions of heaven and hell are not real physical worlds, but projections of the mind realm. They contain a confused and delimited mixture of real world impressions of people, with mythical figures of paradise and monsters, with the heavenly host falling somewhere between.

By contrast, there is another major theme that comes out of Eastern mysticism that is considerably different. The idea goes as follows: In some sense, we, as conscious perceivers caught in the mortal coil of a physical body, are also in some sense manifestations of the cosmic mind, and if we enter into deep meditation we can learn to become one with the Atman, or the Buddha mind, or the Upanishadic 'Self' - the universal seed spark within all sentient beings. In some sense we than become avatars of the one cosmic 'Self', just as the Hindu Gods and Goddesses are in some sense archetypes of the existential condition.

This comes closer to being a valid exploration of the conscious condition and has some root insights, but like the monotheistic myth, it suffers from intractable contradictions, including the notion that nature is merely a delusory gross manifestation subservient to mind, in an overarching moral imperative which causes lustful humans to be reincarnated as animals and vice-versa, and the entire natural world becomes relegated to being merely a cyclical process to refine the (human) ego to the vacuous purpose of attaining oneness with the void and thus escaping eternal suffering. The is a mind dominant fallacy that fails to respect that the diversity of nature, far from being a mere illusion, generates the entire physical basis for our conscious existence.

The meditative quest for 'enlightenment' through becoming one with the cosmic "self" or void

It also introduces the notion of karma in everyday affairs, suggesting that you might end up suffering a nasty accident, or catching a disfiguring disease as a punishment for your egotistical bad actions. This again is a moral causality that clearly runs counter to the needs of life to survive uncluttered by a fantastic causality that runs counter to survival of the organism and the evolution of its genes even when expressed in predatory and parasitic behavior - for carnivores to ruthlessly hunt and kill and even sometimes to torture their prey in honing their hunting prowess.

There are of course many other descriptions of sentient existence, spanning the creation myths of diverse cultures and the shamanistic and prophetic experiences of their various medicine men, diviners, seers, mystics and visionaries, each with their own stories to tell of the vision quest of conscious existence and its relationship with the slings and arrows of outrageous fortune in the world at large.

At best, these become first person accounts of personal experiences and mental voyages, which take us into the territory of the sensitivity of consciousness to apparently supernatural, or paranormal influences, in which one might sense the death of a relative, encounter unpredictable coincidences of fate, or dream prophetic dreams which later appear to come true in real life. At worst they become distracting and delusional fantasies that gain all the features of superstitious beliefs, and cargo cult like mystification.

Where is the Consistency in the Visionary Theatre?

This leads to a basic question. If there is a collective conscious reality out there, shouldn't it be reflected in some way in our mental condition, in our inner meditation and reverie and in our prevailing collective beliefs? If there is 'life' out there in the conscious realm, why are our descriptions of it so idiosyncratic, conflicting and contradictory?

There are a host of reasons for this, some practical and biological, to do with brain function, and others to do with cultural imperatives. We need to take stock of all of these before coming to a synthesis of how we might approach the question of consciousness.

Some mental constructions, such as heaven and hell moral fantasies, are culturally derived from the strong influence major religions have as forces shaping the moral destiny of a culture, quite independently of, and in obvious contradiction to, their truth as a description of the transcendent. Both Deuteronomy and much of the Qur'an deals with unabashedly worldly moral and legal issues, in particular the desire of men to have reproductive control over their women folk and to set them in a partially subservient relationship, as well as driving the formation of powerful large societies of believers, who can gain dominance over perceived enemies and infidels.

Other mental phenomena arise as a reflection of the needs of brain function biologically to compensate for the pressures of daily life and its potential threats to existence and survival. Dreaming remains an enigmatic source of many prophetic and visionary experiences. The nature and rich hallucinogenic content of dreaming remains mysterious, despite extensive scientific investigation. Virtually all of us have had dreams whose richness and power appear every bit as real as waking life experiences, although often much more bizarre, and indeed the only way to subjectively distinguish dream and reality is often a tortuous set of reality checks, such as turning off the light switch and finding the room is still illuminated. Otherwise dreams can seem every bit as real as daily life.

Clearly sleep and the rich REM phases of dreaming have something to do with processing of waking events, either setting down long-term memories, or responding to existential crises which might affect our chances of future survival, but none of this explains the rich, unpredictable and completely bizarre experiential content of many dreams. Dreaming teaches us that almost anything that could be synthesized in the Cartesian theatre of consciousness can appear in dreams, from effortless levitation, to being sucked into the mouth of a giant medusa, or being lost on another galaxy or in another universe, with no clue as to how to find the way back to Ixtlan, or the real world we are familiar with.

Likewise, various drug and plant induced psychedelic states can manifest visions it would be impossible to experience in the physical world at large. These include kaleidoscopic synesthesias, as well as visionary scenes, sensations of going beyond one's bodily confines, as well as a feeling of conscious interplay with the natural world and its psychic subtleties.

Huichol *nierika* or cosmic portal through which the voyager can pass during a peyote vision quest

Again these may be a function of altered brain dynamics, so that one becomes able to perceive in conscious form the dynamic modulations across the cortex induced by these agents, and some of the processes by which the senses are synthesized in consciousness. Thus cultures that use psychedelic species as sacrament tend to explain their existential cosmology in terms of visionary portals or doorways through which one is transported to another reality by the sacramental experience.

Similar considerations apply to a variety of forms of meditation and contemplation, which may also involve stopping the internal dialogue by mantras and/or involve complex visualizations of mandalas not dissimilar in kind to psychedelic kaleidoscopic visions.

These techniques also extend to sensory and/or physical deprivation and so called near death experiences in which people see their life flashing before their eyes, and may claim to meet a luminous entity which is at once themselves and at the same time the 'cosmic mind' meeting them. Again this may be real, but it may also be explained as the product of the extreme, yet living brain state the person was in. Likewise some people report journeys out of the body, which are probably a form of hypnagogic trance on the border of sleep, as they show similarities to levitating dreams, except that they appear to be in the real world environment of the observer, rather than a fantastic dream scene.

Subjective Consciousness and the Objective Brain

One of the ways science tries to explore consciousness is to do experiments eliciting certain brain states while a person is having their brain scanned, either electrically by EEG or MEG encephalograms or physiologically by fMRI or PET, which use magnetic resonance or radioactive scintillation imaging to picture changes in blood flow or nutrient consumption in specific areas. These tend to show what kinds of brain activity or activation are associated with certain kinds of conscious perception, thought or emotion. It can then become possible to see how changing brain activity parallels changes in conscious perception and it can lead to some general hypotheses about how the brain might generate conscious experiences.

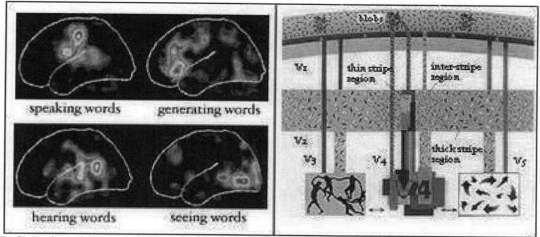

(Left) Brain activity associated with language and (right) local parallel processing of color and motion in vision. Although brain scanning has made it possible to associate specific regions of the cortex with specific aspects of conscious thought and experience, these are just correspondences between biological brain states and perceived conscious events. We still have no idea how the brain actually generates subjective consciousness.

For example the gamma frequency band of the EEG has been suggested to be the excitations the brain uses in active conscious processing and it has been suggested that those networks which rise and fall together 'in phase' constitute conscious processes when they tie together various regions of the cortex into a consistent global dynamical system, by contrast with local processing, which is believed to be unconscious or subconscious.

However these sorts of investigation leave unanswered how the brain makes these global excitations into the internal model of reality which we experience subjectively and identify with the real world around us, or indeed how or why subjective consciousness exists in addition to the computational capacity of the brain as a neuro-system. Because no simple chemical explanation seems to have the right existential status to deal with subjective experience, the problem may need to be solved by examining more exotic physics in the brain, such as quantum entanglement, which might lead to new forms of physical interaction which might solve the problem of existential subjectivity.

Subjective Consciousness as the Existential Complement of the Physical Universe

Critical to the investigation of consciousness is that it is existentially completely different from the objective physical world description, being experienced directly only by the subject, and not being subject to the same criteria of replicability a physical world experiment has. Also the observer cannot control their consciousness objectively in the same manner a physical experimentalist can their equipment, because any attempt to change consciousness carries the observer into a new conscious situation as well. In this respect the exploration of consciousness has similarities to quantum measurement. This renders all forms of introspection made as if we are looking at consciousness objectively, completely, or partially invalid.

It also means that attempts to imagine or model subjective consciousness, or the mental realm, based on objective concepts derived from the physical universe, are invalid because the fundamental properties of the subjective and objective realms are complementary, as opposed to identical, through the symmetry breaking between mind and body. While the physical universe is a process of wave-particle complementarity, in which particulate matter is divisible into real world objects, mind is 'indivisible' in the manner of a 'wave complement' in the sense that it remains the integral field of view embracing all perceivable phenomena, continuous, or discrete. Likewise it is participatory and private in a way which renders objective investigation inoperative.

The exploration of consciousness is thus not the same kind of process as that of the physical world. It is a journey, not a destination. It is a subject experiencing, not an object of investigation. Thus it is not appropriate to try to 'examine' consciousness in the manner of an observation of the real world, but rather exploring it is a 'trip', as the first LSD users, and the sacred mushroom shamaness Maria Sabina, alike have put it, which is where the vision quest of shamanism also takes its journey.

One very positive feature of sacramental shamanism is that it is a visionary experience that can in principle be entered into by anyone in the first person, removing all the disconnections, confabulations and mystifications between the religious follower and the numinous mysterium tremendum that occur with religions governed by gurus, priests, bishops, ayatollahs and muftis, which, rather than being an exploration of the numinous, lead to corrupt religious hierarchies espousing doctrines calculated to preserve their own hegemony.

(Right) San painting of the healing or trance dance Lonyana Rock Kwazulu-Natal. Shamanic trance dancing (centre) in which each participant can witness the world beyond the real world goes back to the emergence of human culture. (Left) San use of *dagga* or cannabis smoking from a hole in the ground. Other ancient pygmy forest cultures utilize the hallucinogenic *iboga* plant.

However, like the previous attempts to understand whether consciousness has any absolute collective nature, we need to remain cautious about the products of psychedelic vision, because these have also led to their fair share of frankly delusional and occasionally violent notions and no definitive conscious cosmology has emerged from many centuries of cultural use of hallucinogens. Nevertheless they are pivotal natural catalysts in the empirical exploration of subjective consciousness.

The Evolutionary Foundation of Subjective Consciousness

To better understand consciousness and the limitations on any speculative ideas of the cosmic

Journal of Consciousness Exploration & Research | December 2010 | Vol. 1 | Issue 9 | pp. 129-140
King, C. *Cutting through the Enigma of Consciousness*

137

conscious connection, we need to ponder how consciousness came about through evolution, and the evolution of the brain.

A likely explanation is that consciousness is an indirect manifestation of the chaotic excitations we see in the electroencephalogram and that it arose in evolution as an offshoot of chaotic excitability in single-celled eucaryotes, which would have provided a multi-sense organ, through the sensitivity to perturbations chaotic excitation provides. A chaotically excitable cell would thus become sensitive to all forms of quantum perturbation of the cell membrane including those of primitive vision, audition and olfaction, as well as electric fields in the medium.

The evolutionary idea of consciousness is that this excitability aided the organism in anticipating threats to its survival, as multi-celled organisms evolved from simple nerve nets, as in hydra, to central nervous systems. Notably many of the critical neurotransmitters involved in changes in consciousness in humans are spread widely across the metazoa down as far as the slime mold, and indeed have distinct psychotropic effects, for example on the web building behavior of spiders.

A critical aspect of this is the idea that such excitations aided anticipating future threats to survival suggesting consciousness is integrally coupled with the notion of free will, or intentional will, which forms another paradox about human activity and existence. All of us feel we have a basic autonomy of choice over our actions and indeed all the provisions of the law, as well as all moral and ethical precepts, revolve around the notion of personal accountability that we can understand the consequences of our actions and can exercise personal control over our affairs.

However this leads to the notion of free will, which appears to be in frank conflict with the idea that our behavior is purely and simply a product of our brain state and its neuro-chemistry and that the notion that we have any purely conscious control over our physical brain states is a delusion. However this need not be true if the brain itself uses exotic quantum physics involving uncertainty in generating consciousness and in the sensitive transitions from chaos to order that may accompany insight learning and decision-making.

Central to an accurate description of subjective consciousness in the universe is the fact that it is, so far as we know, exclusively a product and property of the living biota. In fact the brain forms the most complete interaction of the four fundamental forces of nature in global interaction. There is nowhere else in the universe, from black holes, to dark matter, or the center of stars, that we can plausibly expect to find the physical support for subjective consciousness that we find in the brain of humans, and by extrapolation those of other organisms which possess chaotically excitable brains.

This means that religions posing God as an external agent consciously interacting with humanity, in lieu of humanity's own direct interaction with existential consciousness through our brains, is a fundamental dislocation of reality, removing the direct responsibility we have in participating in consciousness decision-making in our own brains and in taking responsibility for the effects of our actions on the planet, transferring it instead to a physically unrealizable contrivance, in which we become trapped in a moral causality, at the same time passing personal responsibility for our critical decisions on to the will of God.

Even if God is posed as an entity beyond space-time and the universe, the reality is that it is consciousness itself which forms that natural complement to the physical world description. As

Indian philosophy, the Tantric origin and Taoist cosmology put it, the cosmos is a complementarity between subjective and objective reality. Thus the conscious mind, which is the only veridical avenue we have to experience the world around us, may have a cosmological status complementary to the physical universe, despite being manifest in physical terms merely as the excitations of our fragile biological brains.

This symmetry-broken complementarity between the diverse natures of consciousness and matter - personified in the dance of Shiva as observer and Shakti as phenomena - is endlessly reflected in other symmetry-broken complementarities, between wave and particle and boson and fermion in physics, and female and male in biology, something we have termed **the cosmology of sexual paradox.**

While we are standing today, with the benefits of brain science, combined with traditional contemplative techniques, and a diverse array of psychotropic substances, at the threshold of a great exploration of consciousness, which may be the cosmological free lunch the universe is destined to achieve over space-time, we need to realize that many preconceived notions of the purpose of consciousness, or collective consciousness cannot coexist with life as we now know it to be.

For example, it is reasonable, however far-fetched it might seem, to imagine that consciousness might give us access to a form of super-causal quantum future-anticipation which might complement computational brain function to aid survival, but it is not reasonable to suggest that consciousness is there to make us subject to a moral conscience defying evolution's capacity to fill all viable niches, nor to engage in psychic materialism - subjecting conscious experience, by analogy, to constructions derived from the objective world, except in so far as these might be realized in brain function.

Consciousness Arises from the Survival of Natural Life

This brings us full circle to the ultimate questions and quest of conscious exploration. Why are we conscious? What is the meaning in conscious existence? Is there any connection out there with the cosmic mind or any other form of extra-corporeal dis-incarnate from of consciousness?

One thing that is essential to this exploration is that life is sufficient unto itself as it stands without needing either the notion of an after-life or some connection of cosmic consciousness to justify it. We got here because the life force is forthcoming of itself. Although people vary and some people experience depression, partly as a result of genetic variations in brain chemistry, we exist at all only because the web of life has kept an unbroken chain all the way from when the first cells emerged. Life is therefore ultimately productive of itself and is worthwhile simply because it is. The fact that sentient life is also capable of being conscious of itself is a bonus which gives us the capacity to wonder, but it is invalid to turn the tables on life by requiring an after life in heaven to justify the mortal coil.

The key to this is that we are not alone as conscious human beings. Although we have an evolutionary proclivity to procreate and reproduce our genes so that the generations of life continue, we all come to understand that our conscious existence is finite and bounded by our physical demise.

Nevertheless we don't possess our consciousness but are simply conscious of the world and of

Journal of Consciousness Exploration & Research | December 2010 | Vol. 1 | Issue 9 | pp. 129-140
King, C. *Cutting through the Enigma of Consciousness*

139

ourselves. This consciousness is in a fundamental sense a cosmological attribute which we manifest and which is manifest in each and every one of us in various ways due to individual difference but to a great extent consciousness is shared and in common.

This is reflected in the abundance of so-called 'mirror neurons' in the brain which ensure that we are able to consciously experience situations the way others experience them and even feel another's pain. It is also reflected in the oneness that comes from sexual relationship, which is life's antidote to the mortality of the sexual being, in the procreative process and the family.

Because we are mortal, caring is real, not just for the sexual beloved, offspring and kin, but for all mortal beings. Although some people may be violent, psychopathic or selfish, because we are all going to die one day and can't take our possessions with us, the reality of caring for others is what makes both the world, and our consciousness of it, real and worthwhile for each of us. We also leave behind us our humor, art, music and the products of our ingenuity and toil, so the more we contribute to the welfare of the world as a whole, to make it a richer and better experience for all, the better we will feel about life, death and mortality.

This leads us to another fallacy promulgated by traditional religions, which is that the real world is somehow just a flawed secondary realm and that the real existence that makes it all worthwhile is in the after-life. This in turn brings about a sense of futility that if we are going to eventually die, the whole material quest is meaningless dust to dust and ashes to ashes. This is a false description of reality because life is not made worthwhile only because it is eternal, since the web of life is immortal over the vast epochs the planetary environment remains hospitable to life and we each share resonance with conscious existence. We need to keep a perspective of consciousness as a process occurring in space-time, in which the universe is becoming aware of itself through us becoming aware of ourselves during our sentient existence.

Telling Stories Round the Camp Fire

Sentient life is an open-ended awareness whose reality is maintained through the future passage of the ensuing generations of conscious life, so we need to respect preserving the robust fecundity of the planet and its living diversity as a primary task in furthering this quest, in contrast to the linear scorched-Earth eschatologies of monotheistic religion, which risk planetary catastrophe.

Even if the earth is finally vaporized as the sun becomes a red giant and all life is extinguished the conscious quest in the all-embracing envelope of space-time was still the discovery process the universe was able to make to know itself in the alpha-to-omega of all reality space-time is.

In some of my more mystical moments on natural sacraments, I have experienced cosmic consciousness as the bundle of life, as if we as incarnate mortal beings are in eternal communion with all conscious life throughout the universe, from beginning to end, and that when we can for a moment escape the knot in the bundle which our individual survival and ego hold tight and loosen the fibers, we too become one with the cosmic mind.

However, rather than this becoming a description of consciousness, the way for us to move forward is to experience such potentialities for ourselves and keep the description process to personal anecdotes we tell, as conscious participants in the unfolding history of the universe, rather than setting it in stone, as some kind of objective description of how things are out there beyond our personal experience. This is the way people have told stories round the camp fire for the first 100,000 years of human emergence and it serves us well in the electronic age to keep the covenant with the ongoing flow of consciousness between us and among us all to celebrate it personally in our concourse together.

Two very different TOEs attempting to integrate the forces of nature illustrate the intrinsic complexity of the cosmology of the physical universe. If cosmological attributes of subjective Consciousness are a basis of how the brain generates mind, they may have an even more complex basis. This is not to suggest that the model would be like the physical TOE if it exists but that it might have complementary attributes to it.

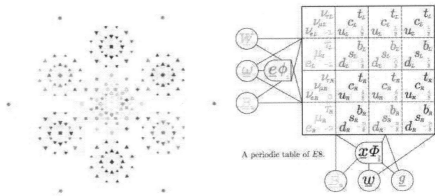

A periodic table of *E*8.

The Intrinsic Complexity of Consciousness and the Ultimate Theory of Everything

But there is another critical aspect to the nature of consciousness which is akin to the difficulty of discovering the theory of everything for the universe, and is so precisely because the conscious brain is the ultimate expression of the four forces of nature derived from the symmetry-breaking of

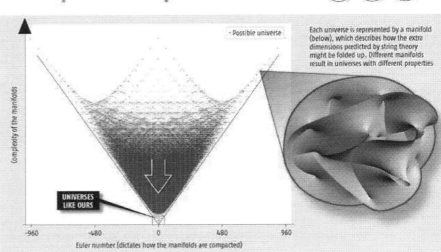

Each universe is represented by a manifold (below), which describes how the extra dimensions predicted by string theory might be folded up. Different manifolds result in universes with different properties

UNIVERSES LIKE OURS

Euler number (dictates how the manifolds are compacted)

the theory of everything. It requires all the forces acting in order of their symmetry-breaking energies to develop molecular matter, and their most complete complex interactive expression we know of is in the human brain.

If the brain uses exotic properties of physics, embracing quantum entanglement in brain states in generating subjective consciousness as part of anticipating future risks to survival, these would be the ultimate interactive structures generated by the symmetry-breaking of the forces of nature in the universe.

Understanding consciousness would then place it as the hard problem complementing the theory of everything, which would require at least as much ingenuity to resolve and therefore cannot be underestimated in the surprises we may find within it.

Essay

Human Consciousness and Selfhood: Potential Underpinnings and Compatibility with Artificial Complex Systems

David Sahner[*]

ABSTRACT

A broadly influenced theory of consciousness and selfhood is presented, followed by a discussion of crucial incompatibilities between human consciousness instantiated in a living biological system and the limitations of artificial intelligence research that might hope to replicate that form of consciousness. It will be argued that human phenomenal experience is firmly anchored in sensation borne of human flesh within a human cultural milieu, and thus enjoys a privileged status. Other pivotal challenges faced by those in pursuit of human-level artificial intelligence are also presented. Based on these considerations, viewed in aggregate, it is concluded that the achievement of human-level artificial intelligence is highly unlikely, even if the potential realization of some form of machine consciousness in the future cannot be excluded.

Key Words: human consciousness, selfhood, artificial intelligence, complex system, biological system.

INTRODUCTION

The field of Artificial Intelligence (AI), which has captivated scientists and philosophers for decades, has also yielded dramatic accomplishments. More recently, the even grander ambitions of Artificial Life (AL) have gained attention and currency within several quarters. Some might claim that the holy grail of such research consists in the successful replication, at a functional level, of broadly instantiated 'human-level' artificial intelligence (HLAI) and, by implication, consciousness of a human variety and style. Nils Nilsson (2010) has referred to the prize in the realm of AI as 'human-made artifacts with levels of intelligence (in all of its manifestations) equaling and exceeding that of humans.'   Here I will argue that this goal is likely to remain permanently elusive in non-embodied artificial non-biological systems.   More specifically, I will contend that human consciousness is predicated upon the existence of a *human body of virtually infinite metabolic and organizational complexity, and the manner in which that body interacts with the environment and other such human beings in a cultural milieu.* And I will assail the notion that virtual bodies and virtual environments are up to the task of adequately simulating what it is like, at an experiential level, for a human being to live (and die) in a brick and mortar world 'red in tooth and claw.' Machine consciousness may be en route, but I believe that if it is born at some future date it will not be human.

I am neither a philosopher nor a computer scientist, although I am familiar with the invaluable output of key individuals in those fields who have weighed in on consciousness and AI. My vantage point is that of a physician-scientist. To contextualize my comments regarding AI, I will sketch an eclectic theory of selfhood and human consciousness that draws significantly from the work of others, outlining, in the process, concepts that enable a fuller appreciation of my later arguments. My influences are ecumenical and, importantly, in the interests of making this paper as broadly applicable as possible, I weave together the strands of this theory with humility, recognizing the possibility of reasonable permutations that have been propounded by various cognitive scientists and philosophers in the field. The focus of the first part of this paper will reside chiefly in what has been dubbed 'phenomenal' consciousness, and is heavily indebted to the published work of Nicholas

Correspondence: David Sahner, M.D., Aeneas Medical Consulting, LLC. E-mail: davidsahner@yahoo.com

Humphrey. In the second and third sections, I will identify and elaborate upon issues of considerable importance in assessing the feasibility of constructing a computer-based replicate of human consciousness. In particular, the challenges by which AI/AL is faced will be outlined with an emphasis on these practical constraints, which, I feel, inhere primarily in ineluctable links between (1) the phenomenology of human consciousness, (2) the byzantine functional organization of the substrate of which it is composed, (3) the evolutionary environment in which this unique form of consciousness has developed, and (4) the social milieu in which the educable human ontogenetically hews his or her own brand of consciousness and selfhood.   *One may certainly disagree with certain underpinnings of the omnibus theory of consciousness put forth in this paper but, even if one does not agree with all of the particulars of the 'how' of the theory, the 'what' of human consciousness that is described is subjectively undeniable, and the arguments supporting the impracticality of HLAI outlined in the second and third portions of this paper will remain unshaken.*

1. An Omnibus Theory Of Selfhood, Qualia, and Phenomenal Consciousness

Any viable theory of consciousness must address what has been dubbed 'the hard problem' of consciousness.   In brief, how does one account for the rich, unique, and highly personal 'phenomenal' experience of the world in which we live? A compelling theory has been put forth by Humphrey (Seeing Red, 2006), who believes 'sensation' and 'perception' constitute two discrete, albeit usually co-occurring, processes that interact. His theory, marked by a visionary bravura, has, quite unfortunately, not yet attracted the following it deserves.   'Perception,' according to Humphrey, neutrally informs the subject about the objects and events beyond the body. 'Sensation', on the other hand, apprises the subject of the response of the body to various kinds of stimuli, generating, in the process, qualia (i.e., the building blocks of our phenomenal experience – the 'redness' of red, the 'coldness' of cold, and the tonality of a given musical note). As Humphrey describes it, 'what sensation does is to track the subject's personal interaction with the external world – creating the sense each person has of being present and engaged, lending a hereness, a nowness, a me-ness to the experience of the present moment.'

Humphrey has suggested that sensation is linked to evolutionarily more primitive responses, and he likens it to an internalized covert bodily 'action' or physical expression-manqué that, among conscious humans, is monitored recursively in a feedback loop that also serves as the basis for the 'thick' moment of phenomenal consciousness within which we live. The activity occurring in both the 'perception-generating' and 'sensation-generating' pathways originates with the registration of energy at the sensory organs. The nervous system response is subsequently split into two, initially independent, channels. The familiar qualitative experience of sensation is produced by the 'sensation' channel.   The other channel, the perceptual one, does not generate any qualitative experience. Normally these two channels respond to stimuli at the same time so that the total experience is a unified whole where non-sensory behavioral competence toward the world is 'clothed', so to speak, in qualitative sensory experiences (qualia). As an apt example of this union, one can cite reflexive withdrawal in response to pain that is triggered by the neutral 'perception' of pain, a perception that is cloaked by the 'painfulness' of this noxious stimulus as an experienced human 'sensation.'

That the two processes described above (perception and sensation) may take place independently has been documented, at least preliminarily, by experimental observation. For example, patients with 'blindsight' appear to have the residual capacity to 'perceive' and appropriately act upon visual stimuli that remain completely opaque to the conscious mind. Conversely, sensation may take place in the absence of 'perceived' external input (e.g., hypnagogic hallucinations, hallucinogen-induced experiences, psychosis, and phantom limb pain).   Humphrey's position that unembellished 'perception' and vibrant conscious 'sensation' follow two distinct avenues within the nervous system is supported by other scientific evidence, including cases of metamorphosia and experimental results of studies that have evaluated sensory substitution, and, more impressively, the phenomenon of sensory mislocation (Armel and Ramachandran, 2003).

Qualia, which form the raw substrate of sensation, possess dimensions that are not directly determined by the real-time perceptual input with which they appear to be allied. For example, a particular suite of 'sensations,' as defined strictly above, and the perceptual knowledge (of what is 'out there') to which these sensations are yoked, may be colored by historically similar sensations that are involuntarily elicited in a manner made famous by Proust's 'madeleine passage' in Swann's Way. Furthermore, the recursive 'monitoring' that serves as the basis for sensation apparently incorporates emotional centers in the brain. Thus, qualia (and phenomenological experience) are affect-laden. The emotional valences of qualia, and the integrated qualia-laden mental constructs that constitute human experience of the world, color the ensuing propositional components of experience (i.e., opinions and beliefs) that subsequently come into being.

Humphrey presents a plausible evolutionary sequence of events underlying the origin and purpose of consciously experienced qualia. This will not be recapitulated here. In brief, there is good reason to believe qualia are not adventitious epiphenomena. In medical and scientific research, one of the best ways to ascertain the function of a molecule or the role played by a given anatomical structure is to study the outlines of its absence. 'Knock-out' mice, lacking a specific gene, provide insights into the biological actions of the protein encoded for by that gene. Similarly, the limitations of stroke victims illuminate the parts played by various parts of the neuraxis. Perhaps the best example of life in the absence of a given class of qualia – without an attendant loss of perception – is embodied in the experience of patients with cortical blindness who retain 'blindsight.' Blindsight was first described scientifically in a monkey in the early 1970's. Since then, there have been a number of reported human cases although the phenomenon remains rare. The clinical observations in these patients are utterly counter-intuitive and, in truth, mind-boggling. They adamantly insist they have no visual conscious experience within the affected portion of the visual field. Yet, on testing, they reproducibly demonstrate an ability to interact appropriately with objects in that field and actually describe visual properties of those objects. In short, they visually perceive but do not sense these objects. In a poignant example of blindsight that highlights the prodigious importance of 'sensation' (as distinguished from 'perception'), Humphrey provides the case study of H.D.:

> 'H.D., a 27-year-old woman, was brought to London from Iran in 1972 to have an operation to remove cataracts from her eyes . . . [When] I was introduced to her, several months after the operation, I found her in a state of great despair. She was convinced the operation had been a complete failure . . . I decided to try some of the same things [I had tried in cultivating blindsight in a prior experiment] . . . I took her out to 'see' the sights of London. We walked the streets and parks, while I held her hand and described what was before her. And soon enough it became clear – to her as well as to me – that she did indeed have a capacity for vision of which she was not aware. She could point to a pigeon on the square, she could reach for a flower, she could step up when she came to a curb . . . She had heard so many accounts, stories, poetry about the wonders of vision. Yet now, here she was, with part of her dream come true, and she simply could not feel it . . . H.D was desperately disappointed, almost suicidal. With great courage, she finally took back control of her situation – by putting on her dark glasses again, taking up her white cane, and going back to her former status of being conventionally blind.' (Seeing Red, page 69).

Elsewhere, Humphrey, I think, identifies the wellspring of H.D.'s despondency, although he doesn't specifically make the connection with this specific case. In elucidating his model, he makes a cogent argument for how the recursive neural circuitry that theoretically underlies qualia (e.g., the phenomenal component of vision that H.D. lacked) presumably provides the neurological substrate for the 'thick moment' of visual consciousness, a consciousness that ties in to selfhood as noted by Humphrey. In short, H.D. was unintentionally teased by the results of her operation in a most horrendous way, much as in the myth of Tantalus, having been brought only to a point shy of visual consciousness – with the attendant dimensional expansion of selfhood that such consciousness would have provided. Here we see, then, an obvious role for qualia. Qualia matter to us because it is their business to matter. There are potential evolutionary advantages to having a rich mental life, not the least of which is the desire to prolong it for as long as possible. As Humphrey observes, a self that has the thick moment of phenomenal consciousness at its core becomes more interested in survival, and this internal make-up may instill added respect for the lives of others who share the capacity for such a conscious life. Phenomenal experience also facilitates the social interaction upon which our (thus far) successful human civilization is based. In particular, empathetic mirroring, perhaps made possible by recently discovered 'mirror neurons,' constitutes a crucial cohesion factor in human culture that appears to be evolutionarily adaptive.

Profound connections and interdependencies appear to exist among the concepts of human 'qualia,' 'emergent mental complexity,' human language, and fitness. Through the acquisition of language, humans have acquired the ability to more accurately describe nuanced qualia-laden mental states (and, by extension, empirical observations of the world), thereby increasing the efficiency of social interaction and knowledge acquisition. The latter confers an advantage through the creation of a burgeoning population of exchangeable ideas, many of which are adaptive. This shared cultural knowledge, albeit of relatively recent vintage, has contributed to our dominance as a species on this planet. Admittedly, written alphabets are only several thousand years old (Dehaene, 2009). The age and genesis of spoken language has been debated, and a detailed review of this topic is beyond the scope of this article. Based, however, on the contention that language facilitates collaborative learning and thereby disallows cultural stasis, William Rowe believes that the archeological record of human tools and artifacts suggests the onset of spoken language occurred approximately 50,000 to 100,000 years ago, a figure similar to that proposed by Noble and Davidson (1996), with a progressive increase in granularity and complexity that has ensued over time since then. Irrespective of the precise chronological age of human language, its cultural evolution is likely to have been tightly yoked to the ontogeny of modern human reflective consciousness and theory of mind. Language has obviously changed the rules of the evolutionary game. In essence, language allows us to symbolically attend to complex qualia-laden experiences and our own internal states, a miraculous capacity poignantly described by Helen Keller over 100 years ago, when she realized, in a minor epiphany, that 'w-a-t-e-r,' as written in tactile letters by her teacher, could represent ' water' as it is phenomenally experienced:

'I stood still, my whole attention fixed upon the motions of her fingers. Suddenly I felt a misty consciousness as of something forgotten--a thrill of returning thought; and somehow the mystery of language was revealed to me. I knew then that 'w-a-t-e-r' meant the wonderful cool something that was flowing over my hand. That living word awakened my soul, gave it light, hope, joy, set it free! There were barriers still, it is true, but barriers that could in time be swept away' (Keller, 1905)

Are linguistically defined qualia-laden mental constructs and internally sensed 'mental states' examples of 'emergent complexity'? That would seem to be the case, but there is no need to believe this opens the door to Cartesian dualism. The existence of discrete 'non-physical' soul-stuff is, at least, one challenge with which AI research need not be encumbered. If one subscribes to non-radical emergentism, which implies the existence of *nested* hierarchies of complexity within emergent systems (Feinberg 2010), it is not necessary to posit the existence of mystical or 'downward' causality originating only at higher strata (e.g., within the realm of human consciousness) that is not directly grounded in the laws and properties of the more fundamental and less abstract workings of the human brain. What likely separates higher-order mental events or symbolically labeled qualia-laden mental constructs from the goings-on within lower layers of the hierarchy is merely pattern recognition by self-reflective creatures such as ourselves. In other words, as suggested by Jaak Panksepp (2007), affective awareness or pre-linguistic phenomenal consciousness, rooted in primitive regions of the brain, serves as the substrate for neocortical parsing of the experienced world into language. Iain McGilchrist (2009) has argued, quite cogently, that the right hemisphere of the brain is also essential to the formation of a holistic and contextually sensitive human impression of the external world based upon heteromodal integration of input from various sense organs; this global pre-reflective element of consciousness serves as a vital prerequisite to the analytic and linguistically rendered left hemisphere representation of reality. In any case, we have created our own lexicon to describe these subjective higher-order patterns and the manner in which they influence each other, but there is no true 'downward causality' that springs from laws that can only be embodied in the higher strata. Our qualia and complex mental states are recognized by us as having certain properties (e.g., a typical etiology, a certain tendency to lead to particular behaviors, a given emotional valence, etc.). These properties are learned and deposited in memory, thereby populating a human lexicon and manner of speaking that is deeply steeped in metaphor which, in turn, is entrenched in the language of a physical *environment as sensed by the human body*. We use these mental constructs to increase the efficiency of communication and to aid in future prediction. It is hard to deny the indispensability of reasonable 'prediction-enabling short-cuts' based on the language of intentionality (e.g., beliefs and desires) – as imperfect as the intentional stance may be at times. Our definitions of mental types and subtypes may be somewhat nebulous on occasion, but they clearly have utility in our syntheses of outward 'cause and effect,' the knowledge of which is transmitted through communal sharing (e.g.,, interpersonal communication). As Nietzsche observed in the nineteenth century, human communication paralleled the development of reflective consciousness/selfhood. Daniel Dennett has pointed out that the intentional stance has adaptive advantages in the sense that human intentional constructs (e.g., wishes, desires) make us better and more efficient future-predictors (Dennett, 1981). The thinking outlined above is entirely consistent with this understanding.

Intuitively, human self-consciousness would seem to consist of a 'higher order' awareness, ephemerally 'experienced' by a transient amalgam of neurochemical processes. Whether or not this represents left hemisphere 'inspection' of right hemisphere activities, as Iain McGilchrist (2009) has suggested, may be debated, but higher order perception/representation (HOP/ HOR) theories of consciousness are well-described. As defined succinctly by Lycan (2004), such theories claim that a state fulfills the definition of a conscious state if and only if 'it is a mental state whose subject is aware of being in it.' I will argue that *human* higher-order awareness is inextricably tied to qualia (and qualia-laden mental constructs) linked to the input of *human* sense organs processed by a *human* brain. Ironically, the instantaneous awareness of a *seemingly durable* self is immediately consigned to oblivion, only to be replaced by a descendent of sorts. The 'enduring self' is a convenient, and, in human culture, pragmatic abstraction, but, as Susan Blackmore has eloquently contended, it is an illusion. An agent partially molds, and is molded by his or her future – but remains subject to the whims of randomness inherent in the environment, which may or may not reinforce specific behavioral repertoires. Our neurophysiologic substrate changes continually, and so do 'we.' These alterations reflect both 'software updates' (and I use this term very loosely) and ongoing hardware modification (and, ultimately, collapse). Although, in a hierarchically organized and multi-tiered nervous system, the line between software and hardware is blurry, it is easy to conceive of neuronal circuitry and the beaten paths within that circuitry of which the self is woven. Probabilistic 'bioprograms' and plastic neuronal circuits in continuous flux, that tend to conduce to one sort of behavioral response over another are dynamic for several reasons, including the insinuation of horizontally transmitted cultural constructs, the level of reinforcement pursuant to each instance of the behavioral execution, and physical renewal/change within the body, including the development of new excitatory or inhibitory synaptic connections based upon experience.

In short, the self is an illusion engendered by cultural consensus which is made possible (and deeply fortified) by human self-reflective capabilities, autobiographical memory, and our competence within the realm of future prediction and, more particularly, future creation. The root of this adaptive and highly functional deception lies within an instinct, namely, the drive toward self-preservation in the face of adversity or threat, which is deeply ingrained. Why species that evince such instinctual behavior would have been more likely to thrive in a Darwinian landscape 'red in tooth and claw' is too obvious to dwell upon. Further evolutionary layers that have been superimposed upon this core kernel of 'the self that must persevere' have only served to reify the sense of self as a seemingly objective and durable entity – notwithstanding the illusion of which it truly consists. As refined 'informavores,' to use Dennett's term, bent upon synthesizing ever-improving prognostications of the future, we have acquired an elaborate and highly quantitative assessment of the passage of time, and the manner in which antecedent events must conspire to successfully bring about some result at a specific date that is yet to arrive. We are acutely aware of how our actions must feed into this complex array of sub-processes in order to win 'self-sustaining' rewards. Given such a capacity for memory, planning, and an appreciation of the future, it is quite natural, and indeed expedient, to think of the agent negotiating the river of time as some sort of discrete and durable noumenal entity that sees the ship to port.

Social interactions and cultural beliefs – the sea of cultural constructs in which we are immersed on a daily basis – solidify this sense of a lasting self. One example will suffice here. People are graded on performance, both professionally and personally. A focus on performance, as a species, conveys self-evident adaptive advantages although this preoccupation may be taken to grotesque extremes. In short, people frequently believe 'self-worth' is linked to past accomplishments and failures, for which they are forced to take responsibility even if dramatic supervening events have seen to it that they are, quite clearly, not the people who made those blunders in the past. Many of our culture's most coveted awards recognize *years* of stellar contributions within an arena, even if the prize is specifically and nominally awarded for a given contribution. One person is said to have produced all of these contributions, but, as defined above, this is patently untrue. Such societal values and pressures, essentially glorifying the myth of an enduring self, lead to William Faulkner's 'eternal human verities,' of which pride is a notorious example. In addition, as future predictors – and makers – a powerful basis for the concern we harbor for our subsequent 'incarnations,' even if there is no enduring self, is the knowledge that those later avatars, apart from sharing a good deal of our current physical substrate, will have the clearest recollections of who we are at this 'earlier' moment. They will earnestly care about and relish the remembered experiences of the prior iterations of the self. We don't typically think in such stark terms but all of us know that the appearance, behavior, and accomplishments of our 'ancestors' (i.e., our prior selves) and 'successors' (i.e., our future selves) will color the impressions of 'us' carried by future generations. In essence, given our self-awareness and the knowledge of our own mortality, we collaborate throughout our lives with the many successive embodiments of ourselves (i.e., the fleeting instantiations of the 'I' described above) *to erect virtual 'self monuments.'* How can a machine, even one with the capacity to 'learn,' ever hope to mimic the goings-on within a virtual and malleable instantiation of human consciousness and the successive yet linked embodiments upon which it is reliant over the duration of a human life?

2. Human-Level Artificial Intelligence/Artificial Life: Perils and Pitfalls

Endeavors to reproduce human consciousness within a computer system have been pronounced ill-fated by those would argue that a computer is a formal syntactical system devoid of all semantics (Searle, 1984).   This serves as the basis for Searle's 'Chinese room' argument that has spurred considerable and spirited debate. Without meaning, and without the tethering of internal tokens to such meaning, there can be no consciousness. 'Information' is a *social* construct.   The computer is viewed as a perfect yet insular syntactical island, capable of performing extraordinary computations at rapid speed, but within a programmed environment derelict of meaning.   A Universal Turing machine can be programmed, however, to flawlessly copy the behavior and actions of any formal system. It has been claimed that, for a *truth-preserving* formal system, the 'semantics will take care of itself' (Haugeland 1981). Yet human beings are not always truth-preserving and there may be solid evolutionary reasons for this (McKay and Dennett 2009). Haugeland concedes that truth is not the only important thing – one needs to make sense (as defined by concepts such as 'rationality' and 'conversational cooperativeness'). Haugeland appropriately expresses hesitation when it comes to the question of whether such a system could mirror the 'intelligence manifested in everyday life, not to mention art, invention, and discovery.'

AI/AL faces other major challenges, as outlined in the bullets (**A** through **G**) below:

A. The precise *manner* in which an organism interacts with the external environment defines the 'currency' of that organism's experienced qualia and resultant phenomenal consciousness. A computer can't experience the profound feeling of involuntary facial blushing and warmth that we link to the sensation of embarrassment, and that we know outwardly betrays the knowledge of a fault. A computer has no vascular system, so any 'functional analog' of embarrassment will inadequately equate with the 'full-blooded' human experience. One could cite many other similar examples. In brief, human qualia and qualia-laden higher-order mental constructs would appear to be anchored in human flesh.

B. In humans, qualia, qualia-laden mental constructs, and self-reflective consciousness have evolved and conspired to create deeper *meaning*, fortifying the instinct toward self-preservation and reproductive success. The perishability of organically-grown consciousness serves as the context for meaning, defining why certain experiential phenomena (e.g., pain) are experienced in the manner in which they are experienced, and why certain types of behaviors take up residence in an organism's behavioral repertoire. Computers with 'on and off' switches exist in a sphere entirely different from that of people continually poised at the interface of 'life and death.' Meaning is *born* of these draconian consequences (i.e., personal extinction).   As Searle has perceptively observed, intentionality and understanding are features of evolved biological systems.

C. 'Meaning' for humans, which is a precondition for phenomenally rich experience of the human type, is molded by social interaction and cultural influences in a brick and mortar world. Some human forms of meaning, and, more particularly, the behaviors by which they are attended, are likely to have been selected for at a group or societal level, much as evolution instigates change based on adaptive advantages conferred at a colony level among the hymenoptera (Holldobler and Wilson, 2009).   Cooperation, through highly efficient communication, has been critical to human success. As Charles Whitehead has remarked, 'Of course competition is an important aspect of biology, but without cooperation, there would be no competitors' (Whitehead, 2010). The implications of this observation for AI are glaring.

Without a societal context in which to evolve (and develop meaning), machine AI is severely hamstrung. Virtual AI embodiments operating within a virtual culture would provide only an impoverished imitation of 'what it is like' to experience phenomenal human consciousness as complex and metabolically active embodied agents living and dying in the real world.

D. Profound and likely insurmountable practical challenges would inhere in any effort to artificially reproduce the complex and dynamic functionality of a human nervous system, which has been underestimated.   While HLAI proponents such as Ray Kurzweil do acknowledge the existence of '100 trillion connections' in the brain, the enormous variety of those (mutable) connections is staggering. There exist multiple neural phenotypes, over 100 neurotransmitters (with others that will surely be discovered), and an even wider variety of neurotransmitter receptors. In some cases, synaptic vesicles contain more than one type of neurotransmitter, and even a single neurotransmitter may have excitatory or inhibitory effects based upon local ionic concentrations inside and outside of the cell, and the specific permeability of the ion channel directly or indirectly affected by the binding of that neurotransmitter.   In addition, the binding efficiency of a neurotransmitter to its cognate receptor may be influenced by receptor accessibility within membranes that behave like non-linear complex systems – the composition, organization, and fluidity of which appear to vary in response to incompletely understood factors (Garcia et al., 2002). Receptor desensitization may also be affected in a multifactorial manner. For example, there are at least several different mechanisms underlying reversible NMDA receptor desensitization, and, to complicate matters further, the kinetics of receptor inactivation may be influenced by experimental methodology (Kyrozis et al., 1996). Beyond all of this, molecular signaling within neurons, after neurotransmitter binding, entails an intricate cascade of multiple reactions including generation of second messengers and effector proteins. Changes in the intracellular availability of these downstream messengers/effectors may affect neuronal function.   And none of this is static in a living organism. How can an amalgam of 'integrated circuit work-alikes' precisely mimic the diverse cellular functionality of the composite of all of our neurons, even in a crude sense? The number of degrees of freedom would seem to be impossibly large.

The HLAI advocate might counter this particular argument by stating that all of the above-described complexity is well and good, but the bottom line is that the end product (the 'spike' or action potential) encoding information is an 'all-or-none phenomenon' that is easy enough to mimic based on an integrated analysis of inhibitory and excitatory input to a given 'silicon nerve.' There are two major problems with this stance. First, the underlying changeability and 'finesse' of the process that *determines* whether a firing threshold has been reached is lost and, presumably, that incredible granularity, as outlined in the preceding paragraph, is, *functionally* important. If it were not, then why would so much energy be expended in maintaining it? Second, as Bill Rowe has keenly observed (personal correspondence) 'informatics models require fixed transfer functions that always give . . . a predictable output for a given set of input conditions,' yet these models cannot reasonably accommodate a 'poisson-like' spiking rate that exhibits only 'weak coupling [to the] neural population' of which it is a part. This latter observation may be terrifically important, and is supported by the work of Walter Freeman (2000) who has shown that there is a high degree of chaos/randomness and a low degree of covariance among spiking neurons within a given functional area. *This minimal synchronicity, however, can accommodate amplitude-modulated patterning at a mesoscopic scale.* Stimulation enhances synchrony, leading to *a state transition that is propagated in neural tissue.* This is *not* equivalent to the yes/no firing of a single neuron. During the propagation of a mesoscopic signal, shifting voltage isobars reflect attendant changes in local current. If the latter is a means by which information is conveyed in the human brain, it is dubious that it could be modeled in silico. In the words of Bill Rowe (personal communication), 'with a near-random spiking rate, there can be a causal relationship between activity in the dendrites and axonal firing, but it cannot be computational.' Moreover, how could one mimic any 'functional cooperation' among propagating amplitude-modulated patterns at the mesoscopic level?

To further confound matters, AI advocates must also grapple with incompletely understood glial cell functionality. The glial cell population in the brain is about three times larger than the sum total of all of the neurons in that organ. These cells influence the rate of signal propagation in nerves, synaptic activity and the environmental concentrations of ions which, as described above, may affect the character of a postsynaptic potential (i.e., excitatory versus inhibitory).

Finally, Jaak Panksepp (2007) has suggested that the affective element of consciousness, which is grounded in activity within sub-neocortical territories of the brain (e.g., limbic pathways, midline mesencephalic region), may be mediated by broadly acting neuromodulators such as the neuropeptides. That is, 'emotional states,' a term in vernacular use, may have neurophysiological corollates that are relatively refractory to the type of analysis that seeks to explain cognitive dimensions of consciousness mediated by information channels and computation. The in silico modeling of such 'un-bordered' emotional state influences, not to mention the manner in which the resultant emotional states dialog with cognitive neocortical activity, would represent a prodigious challenge. And neuropeptides are not the only type of neurotransmitters that may modulate the electrical activity of noncontiguous neurons at some distance from the site of release. Endogenously produced nitric oxide appears to share this same property. In a similar vein, remotely synthesized hormones influence central nervous system (CNS) function, as evidence by the glaring CNS manifestations of certain endocrinopathies (e.g., hypothyroidism and Cushing's syndrome).

Given all of the above-described subtleties and potential permutations that affect neuronal function, even if three-dimensional nanotube-based electronics technology one day furnishes an adequate hardware base, who could program a machine to provide a faithful simulacrum of human intelligence and consciousness? One would have to contend with a virtually infinite number of continuous (rather than dichotomous) variables in a state of ongoing flux merely in order to determine whether the instantaneous integrated influence of the system on *one* neuron (including the effects of up to 100,000 potentially *variable* synaptic inputs, *remotely acting* neurotransmitters that presumably follow a concentration gradient, hormones, a changing intracellular milieu, glial cells, etc.) is such that the neuron will spike. Multiply this complexity by 100 billion (the number of neurons in the human brain). And now provide the correct software to broadly mimic human intelligence. Ray Kurzweil (2010) has suggested that a more modest goal be set initially, and that forms of machine intelligence start life as childlike tyros that are subsequently educated 'just as we do with natural intelligence.' For reasons discussed toward the end of this paper, I don't think the latter strategy will yield HLAI.

E. Margaret Boden (1999) has compellingly contended that metabolism in a strong sense should be regarded as necessary for artificial life which, has been argued, is an essential prerequisite, in turn, for cognition. She defines metabolism in the strong sense as 'the use, and budgeting, of energy for bodily construction and maintenance, as well as for behavior,' and further characterizes it as 'a type of material self-organization which . . . involves the autonomous use of matter and energy in building, growing, developing, and maintaining the bodily fabric of a living thing.' Boden states, and quite rightly I believe, that 'virtual creatures . . . would be pale simulations of the real thing' (i.e., a materially embodied metabolically active being).

F. Ned Block has cogently criticized the basic premises of Machine Functionalism, the viability of which is critical to the realization of any attempt to functionally instantiate human consciousness in silico. Block has proffered arguments against functionalism in describing his China-body system, consisting of a functionally organized network composed of a billion individuals communicating with each other in a manner that is equivalent to the operation of the brain of an individual (Block, 1978). This system, although potentially mimicking the hierarchical structure of a conscious mind engaged in complex tasks would be, in zombie-like fashion, devoid of true consciousness. Perhaps more compellingly, Block has pointed out that functionalism is bankrupt at the interface between the organized system and the external world. In particular, definitions of inputs and outputs lead to infinite disjunctions based on species-specific differences in sensory organs and body morphology. Block (1978) has brilliantly suggested that *'the brain itself becomes an essential part of one's input and output devices,'* quoting the efforts of scientists at UCLA studying the capacity of an 'EEG to control machines . . . A subject puts electrodes on his scalp, and thinks an object through a maze.' This technology has been much refined in recent years (Lu et. al., 2009). A recent publication in the New England Journal of Medicine also buttresses this claim (Monti et al., 2010). In the latter study, subjects were able to communicate 'yes' or 'no' answers to questions by deploying specific forms of mental imagery detected by corresponding neuroanatomical signatures visible on functional magnetic resonance imaging of the brain. In short, the brain itself was used as a 'voice box' with no intermediary organs or tissue. *If brains may serve as direct output devices, it may be vain to consider the possibility of a full cross-species characterization of 'inputs and outputs' that can be used to define the 'functional identicalness' of mental states instantiated within different systems.* While simple translations of 'yes' or 'no' outputs, as in the recent experiment above, can be functionally characterized, functional transliterations of more complex and nuanced mental configurations (i.e., how one might feel after having first lost a queen in a chess game with one's father) would seem to be futile. The corresponding 'neuroanatomical signature' or 'output' would be uniquely human. In short, if there exists a type of 'output' that is, in and of itself, strictly defined by the specific anatomical or physical attributes of the system, then it is impossible to 'functionally duplicate' the same output by using a qualitatively different system.

G. As Lakoff and Johnson (1980) have pointed out, metaphorical concepts *govern the manner in which we think* (Lakoff and Johnson, 1980). Metaphor is embedded in the modes in which we, as embodied entities, interact with and experience the world. Symbolic representations of such metaphorical relationships, however, represent profound challenges in silico. Yet

Journal of Consciousness Exploration & Research| December 2010 | Vol. 1 | Issue 9 | pp. 141-155
Sahner, D. *Human Consciousness and Selfhood: Potential Underpinnings and Compatibility with Artificial Complex Systems*

151

without the metaphorical overlay, understanding is compromised or negated.Lacking a first-hand experiential knowledge of human qualia-laden mental constructs, machine intelligence would be at a distinct disadvantage in interpreting metaphorical utterances or text. Other features of the semantics of written and spoken language pose challenges to AI as well. For example, meaning is *context-driven*, and a single word may convey, to a greater or lesser extent, *multiple and simultaneous direct meanings and connotations* that color the manner in which a human utterance is interpreted in a given instance. *Contemporaneous non-verbal behavior and tone also mold meaning* during speech in endlessly complex ways. One-to-one mapping of a word to meaning may be impossible given such complexity and fluidity. In short, without social context and effective communication between agents in the metaphorical language of human beings, a language made real through the existence of the human body and the specific ways in which it interacts with, and senses, the ambient world (and itself, for that matter), there is little hope of recreating human consciousness in silico.

.

Perhaps the most robust counter to some (but not all) of criticisms **A** through **G** leveled above at HLAI/Artificial Life enthusiasts is to be found, either implicitly or explicitly, in various sections of a magisterial work released in 2009 by Melanie Mitchell, entitled *Complexity: A Guided Tour*. This highly ambitious overview adopts a multi-disciplinary approach to complexity (and, by extension, the possibility, in principle, of artificial life) grounded in the evolving fields of chaos, evolution, genetics, computer science and network theory. Although Mitchell admits that there exists no universally agreed upon definition of complexity, or means by which to effectively measure it, her concrete examples illustrate the emergence of what many would intuitively regard as 'complexity' within the fabric of decentralized systems composed of numerous entities, each of which is characterized by relative simplicity. Her most pithy formulation, that complexity implies the existence of 'non-trivial emergent and self-organizing behaviors,' seems quite reasonable (if 'self' is defined in a suitably broad way). Particularly admirable are Mitchell's careful attempts to define certain terms very precisely, such as 'information.' A narrow definition of 'information' is provided but she also boldly states, and very appropriately, that the definition of information *should be tethered to 'meaning.' Of* course, this raises the question: *'Meaning to whom?' In* this vein, and very importantly, she acknowledges that meaning is a 'thorny' issue and that there is no completely palatable explanation for human consciousness. Mitchell links 'meaning' to survival and natural selection. For example, if X is relevant to my well-being or prospects for procreation, then it is meaningful to me. As a general definition, however, that may be applied to the rich tapestry of various human-perceived 'meanings,' this seems to fall far short of the mark. Such meaning often takes forms unrelated to the principles of survival, as evidenced by the multifarious nuances of human language. Meaning *to us, and* the language that we employ to convey that meaning, is woven of human subjective sensation, cognitive overlay, and *cultural dic*tates. The definitions of the vast majority of words in the English language have nothing to do with survival or the perils of extinction even though these concepts are obviously important to us.

Mitchell mentions the artificial life (AL) community's claims to have dismantled arguments against AL based on commonly held beliefs in the inimitability of such life-defining features as autonomy, metabolism, self-reproduction, adaptation, and instincts for survival. In her own book, she reviews genetic algorithms that effectively permit a basic form of 'selection' to occur in silico, the discovery of the self-reproducing automaton, computation within cellular automata, and, even a computer program (Copycat) that is capable of making basic analogies - the key step on the road to metaphor. The experimental observations that are marshaled, which seem to accommodate the notion of artificial life of some sort, at least in principle, are daunting. A detailed review of these specific topics far exceeds the scope of this monograph but, even if one grants that all of the properties listed above can be simulated or reproduced in silico, at least in theory, a fundamental strut of the counter-argument (i.e., against human consciousness-endowed AL) still stands: *Meaning as perceived by a human being is tethered to complex mental states woven of human qualia that are themselves grounded in perceptual and sensory pathways composed of uniquely organic human tissue within an embodied framework. In addition, human meaning is marked by an essential external dimension anchored in human cultural consensus. Therefore, the nuances of human meaning and metaphor, which are anchored in our particular physicality, cannot be perfectly reproduced in silico and, therefore, human consciousness cannot be faithfully duplicated in such an artificial system.* On a more minor note, cellular automata, as described by Mitchell, cannot mimic broadly acting state influences (such as neuropeptides) because each constituent of a cellular automaton is linked only to the cells it abuts within its neighborhood. Network theory, of course, accommodates distant connections, and Mitchell does liken the brain to a network. I suspect it may be theoretically possible to model the influence of neuropeptides within a brain by relying upon a suitably organized network. In such a case, the source (or sources) of such molecules could be considered as 'nodes' with vastly polyvalent 'out-links' to enormous swaths of the system. But then, again, this doesn't address the key objections raised above.

3. What Then Can We Say About Human Consciousness and the Prospects of its Duplication In Silico?

Some of the greatest contemporary thinkers continue to wrestle with the slippery notion of consciousness. At times, they seem to speak to each other at cross-purposes, in part because of the absence of unambiguous case definitions of basic concepts such as 'consciousness' itself. Elements of various theories seem to ring true but, ultimately, efforts to gather compelling empirical evidence in favor of any theory remain stymied by the uniquely subjective and internal experience of consciousness.   Some materialists, such as Daniel Dennett, have concluded that we, as conscious entities, are not in or out of the loop on an instantaneous basis but that, rather, 'we *are* the loop,' and that our 'mental powers [are] smeared out over time' (Dennett, 2003) . With this many would agree, but the devil, of course, is in the details.

Reflective consciousness, meta-awareness, stored representations, beliefs, and learning have evolved over billions of years. At the summit of this evolutionary process sits man, a complex intentional creature capable of learning and sharing the fruits of that learning with conspecifics, contributing to the dominance of the human species on earth. Learning is an adaptive process capable of rapidly and productively influencing behavior in a manner that is superior to evolutionary heuristics (Millikan, 1989). The sharing of knowledge and cultural indoctrination produce clear group benefit. Language represents the basis for that interchange.   In 1887, Nietzsche recognized that consciousness paralleled the capacity for communication.   Adopting an economical 'intentional stance' toward others, in which we rely upon folk psychological terms like beliefs and desires, makes of us better and more efficient future predictors. Attribution of these beliefs and desires to ourselves and others is adaptive – but who is making these attributions? A virtual leader or chief executive officer of sorts is at the helm, but one whose role cannot be usurped by a machine. For the reasons outlined in this paper, beliefs and desires grounded in uniquely human sensations and experiences can't be replicated in silico. The virtual captain of a human vessel essentially pilots a perishable ship. In a world 'red of tooth,' survival and reproduction are the basic criteria for success. Our highly differentiated cells, tissues, and organs comprise bodies bent on survival, bodies belonging to agents possessive of 'high-level intentionality' by *evolutionary default*. Even if adaptive evolution can be

Journal of Consciousness Exploration & Research| December 2010 | Vol. 1 | Issue 9 | pp. 141-155

Sahner, D. Human Consciousness and Selfhood: Potential Underpinnings and Compatibility with Artificial Complex Systems

153

modeled in silico, a gulf would still exist between the qualia-laden complex mental structures of man and the 'inner life' of an autonomous and self-reproducing robot.

Denis Dutton (2009) has argued that the human capacity for the creation and admiration of art may have evolved as a mate-attracting strategy. Art generated instinctually *for such a purpose* could hardly be imitated by a non-reproducing digital computer, no matter how byzantine its software may be. *In short, the reason for the existence of a behavior partially defines that behavior. As such, computers with on-off switches differ radically from sentient beings such as humans, for whom 'on and off' translate into the far more draconian 'dead or alive.' These mortal consequences, and our uniquely human phenomenal experience, help create the semantics of human consciousness. Cultural factors are absolutely pivotal in defining that semantics.* AL efforts may move closer to a realization of 'consciousness' of some sort when artificial systems have evolutionarily definable 'skin in the game,' and a vested interest in the future that *confers meaning.* Based on considerations discussed earlier, the advent of communication between such 'Darwinized' and embodied artificial systems, and the dawn of a corresponding culture, might be expected to parallel the development of a novel type of 'artificial' consciousness, but this remains very speculative. Even so, distinctly *human* sensations, beliefs, memories, and cognition – in short, all the ingredients of the consciousness of a complex biological intentional system such as ours, are borne of flesh-based sensory organs and neurobiological processes that cannot be mirrored precisely in silico.

Ultimately, I believe that Robin Zebrowski (2010) is absolutely correct in claiming that 'embodied' AI may one day be poised to unlock the mystery of the mind. It is only by 'real sensing' as she puts it, through, among other things, mobility and physical interaction with the environment, that an artificial complex system (read: robotic form of AI?) can hope to acquire 'felt experience' (again, her phrase). Obviously, it will take more than a Mars Rover to instantiate consciousness. Zebrowski (2010) astutely observes that promulgators of 'good old-fashioned AI' are wide of the mark in the sense that:

> 'They fail to recognize that the visual system is simply one aspect of the entire body, but even more they fail to recognize that the visual system is itself an *active* (my italics) system, one that pushes back on the world by selecting for attention and imposing some higher-level interpretations on it, and not merely something which receives pictures for processing . . . Vision is not the process of taking merely what is shown to us of the world, internalizing it, and performing some manner of intellectual operation on it . . . .'

The parallels between this insight, and Humphrey's phenomenology, are uncanny. And in keeping with what I have written in this manuscript, she quotes Rodney Brooks in emphasizing the importance of 'the ability to carry out survival related tasks in a dynamic environment [as a] necessary basis for the development of true intelligence.' I would claim that it is fair to substitute the term 'consciousness' for 'intelligence' here. Finally, Zebrowski homes in on the import of human language, metaphor and meaning in defining human understanding. All of this is inextricably bound up in human culture. Similarly, the British psychiatrist Iain McGilchrist (2009) emphasizes the importance of embodiment to human experience, and the manner in which metaphor, grounded in that physicality, creates bona fide human understanding.

Like Robin Zebrowski and Iain McGilchrist, Mitchell Kapor, in arguing that machine intelligence will not have passed the Turing test by the year 2029 (http://www.longbets.org/1), states 'our physicality grounds us and defines our existence in a myriad of ways.' And he correctly points out that our phenomenal experience constitutes 'the basis of agency, memory and identity.' He also cleverly observes that a good deal of human experience and knowledge is 'tacit' and inexpressible. Kapor is absolutely on target here. Ultimately, not even poetry can completely satisfy the goal of perfectly reproducing, in another human being, the precise inner state of the poet. At their core, qualia and, by extension, integrated phenomenal experiences in general, defy 'verbatim' description. Through

Journal of Consciousness Exploration & Research| December 2010 | Vol. 1 | Issue 9 | pp. 141-155

Sahner, D. Human Consciousness and Selfhood: Potential Underpinnings and Compatibility with Artificial Complex Systems

154

prestidigitation, the poet may come close to a faithful verbal simulacrum of phenomenal experience – but metaphor is an evasive tool. If elements of human knowledge, sensation and experience are not amenable to perfect description in language then how, as Kapor remarks, can Ray Kurzweill's program by which a computer is to be endowed with all human knowledge (i.e., through the relentless review of all of human literature) ever to succeed?

Let me take my argument to what I believe is its natural conclusion in a single pithy claim: AI/AL efforts to duplicate human consciousness may succeed in the future, if the exponents of AI build a machine functionally identical to that of the human body in every byzantine detail. Moreover, this body would need to interact with its landscape and other similar agents (i.e., its conspecifics) as we do. But this, then, becomes a trivial tautology since one would have simply built, in essence, a human being – and a human being, by definition, owns a human consciousness. Functionalism, it would seem, may have its practical limitations within the dominion of AI. David Chalmers (2010) admits that the possibility of an artificial 'functional isomorph' of a human brain represents a 'substantive claim.' With this I would fully agree. For the reasons cited earlier in this paper, I don't think such a functional isomorph is possible. Yet, even if it were, more is necessary to accurately engender human consciousness within a machine or artifact. This functional isomorph of the brain would need to be precisely linked to a perishable isomorph of the human body that provides the substrate for qualitatively accurate human qualia and integrated phenomenal experience. Furthermore, this body, in turn, would need to be appropriately enmeshed within the weft of human society (or a functional isomorph of human society) as it exists on earth. I do not have the temerity to believe that this feat of legerdemain can be accomplished by anyone, now or in the future. Although I certainly cannot discount the possibility of some form of machine consciousness at a later date, if it is successfully cultivated it will not be equivalent to human consciousness. A conscious computer would have as much difficulty imagining what it is like to experience the world in the full-blooded manner of a human being as we have in attempting to envision what it is like to be a bat.

Acknowledgements: This manuscript was prepared, in part, in the midst of extensive discussions and correspondence with William Rowe and, as such, includes an amalgam of my own ideas as well as some of his. I am deeply indebted to him for his careful review of this paper, as well as his astute insights, contributions and support. I also thank Daniel Dennett, Melanie Mitchell, Nicholas Humphrey and Andrew Neher for their comments and suggestions.

References

1. Armel, K. and Ramachandran V. (2003). Projecting sensations to external objects: Evidence from Skin Conductance Response. *Proceedings of the Royal Society of London: Biological* 270 1499-1506.
2. Block, N. (1978). Troubles with Functionalism. In: W. Savage (Ed.), *Perception and Cognition: Minnesota Studies in the Philosophy of Science IX* (University of Minnesota Press)
3. Boden, M. (1999). Is Metabolism Necessary? *Brit. J. Phil. Sci.* 50 231-248
4. Chalmers, D. (2010). The Singularity: A Philosophical Analysis. *JCS*, 17, No. 9-10, pp. 7-65.
5. Dehaene, S. (2009). *Reading in the Brain* (Viking)
6. Dennett, D. C. (1981). True Believers: The Intentional Strategy and Why it Works. In: A. Heath (Ed.), *Scientific Explanation: Papers based on Herbert Spencer Lectures given in the University of Oxford* (Oxford University Press)
7. Dennett. D.C. (1991). *Consciousness Explained* (Boston: Brown and Co.)
8. Dutton, D. (2009). *The Art Instinct* (Bloomsbury Press)
9. Feinberg, T. (2010). Why the mind is not a radically emergent feature of the brain. *JCS,, 8*, 123-45
10. Freeman, W. (In Press: September 2000). Characteristics of the synchronization of brain activity imposed by finite conduction velocities in axons. *International Journal of Bifurcation and Chaos* 10: 2307-2322.
11. Garcia, D., Marin, R., Perillo, M. (2002). Stress-induced decrement in the plasticity of the physical properties of chick brain membranes. *Mol Membr Biol., 19,*221-30

Journal of Consciousness Exploration & Research| December 2010 | Vol. 1 | Issue 9 | pp. 141-155   155

Sahner, D. *Human Consciousness and Selfhood: Potential Underpinnings and Compatibility with Artificial Complex Systems*

12. Haugeland J. (1981). Semantic Engines: An Introduction to Mind Design. In: J. Haugeland (Ed.), *Mind Design: Philosophy, Psychology, and Artificial* (MIT Press)

13. Holldobler and Wilson (2009). *The Superorganism* (Norton).

14. Humphrey, N. (2006). *Seeing Red: A Study in Consciousness* (Cambridge and London: Harvard University Press)

15. Kapor, M. and Kurzweil, R. Opposing arguments against (Kapor) and in favor (Kurzweil) of the existence of machine intelligence capable of passing the Turing test by the year 2029. Accessed on the Internet 19-OCT-2010 (http://www.longbets.or/1)

16. Keller, H. (1902, 1903, 1905). *The Story of My Life* (New York: Grosset & Dunlap)

17. Kyrozis A., Albuquerque, C., MacDermott, A. (1996). Ca^{++}-dependent inactivation of NMDA receptors: fast kinetics and high Ca^{++}-sensitivity in rat dorsal horn neurons. *J. Physiol., 495 (Pt 2),* 449-63

18. Lakoff G. and Johnson M. (1980). *Metaphors We Live By* (University of Chicago Press)

19. Lu, S. et al. (2009). Unsupervised brain computer interface based on intersubject information and online adaptation. *Neural Systems and Rehabilitation Engineering, 17,* 135-45.

20. Lycan, W.G. (2004). The Superiority of HOP to HOT. In: R. Gennaro (Ed.), *Higher-Order Theories of Consciousness* (John Benjamins)

21. McGilchrist, I. (2009). *The Master and his Emissary: The Divided Brain and the Making of the Western World* (New Haven and London: Yale University Press)

22. McKay, R. and Dennett, D. (2009). The evolution of misbelief. *Behavioral and Brain Sciences, 32,* 493-561

23. Millikan, R.G. (1989). Biosemantics. *Journal of Philosophy LXXXVI, 6,* 281-97

24. Mitchell, M. (2009). *Complexity: A Guided Tour* (Oxford University Press)

25. Monti M. et al. (2010). Willful modulation of brain activity in disorders of consciousness. *New England Journal of Medicine, 362,* 579-89

26. Nietzsche F., (1882 and 1887). *The Gay Science*

27. Nilsson N. (2010). *The Quest for Artificial Intelligence: A History of Ideas and Achievements* (Cambridge University Press)

28. Noble, W. and Davidson I. (1996). *Human Evolution, Language and Mind: A Psychological and Archaeological Enquiry* (Cambridge University Press)

29. Panksepp, J. (2007). Affective Consciousness. In: M. Velmans and S. Schneider (Ed.), *The Blackwell Companion to Consciousness* (Blackwell)

30. Purves, P., et al. (2008). *Neuroscience, Fourth Edition* (Sunderland, Massachusetts: Sinauer Associates)

31. Searle, J.R. (1984). Can Computers Think? In: *Minds, Brain and Science*, pp.28-41

32. Tennyson, A. "In Memoriam A. H. H." (the poem serving as the source of the phrase "red in tooth and claw")

33. Whitehead, C. (2010). Rethinking Reality (editor's introduction to a special issue of *JCS*). *JCS,* 17, No. 7-8, pp. 7-17.

34. Zebrowski, R. (2010). In Dialogue With the World. *JCS,* 17, No. 7-8, pp. 7-17 (pp. 156-172).

Book Review

David J. Chalmers: *The Character of Consciousness*
Oxford University Press, USA, 2010, 624 pp. ISBN-10: 0195311108

The Character of Consciousness

Peter Hankins[*]

ABSTRACT

It's a good, helpful book; what the content lacks in novelty it makes up in clarity. Chalmers has a persuasive style, and his expositions come across as moderate and sensible (perhaps the reduced epiphenomenalism helps a bit). It's surprising that the denial of materialism (surely the dominant view of our time) can seem so common sense.

Key Words: consciousness, character, David Chalmers, materialism, dualism, hard problem, neural correlates of consciousness.

The Conscious Mind was something of a blockbuster, as serious philosophical works go, so a big new book from David Chalmers is undoubtedly an event. Anyone who might have been hoping for a recantation of his earlier views, or a radical new direction, will be disappointed – Chalmers himself says he is a little less enthusiastic about epiphenomenalism and a little more about a central place for intentionality, and that's about it. *The Character of Consciousness* is partly a consolidation, bringing together pieces published separately over the last few years; but the restatement does also show how his views have developed, broadening into new areas while clarifying and reinforcing others.

What are those views? Chalmers begins by setting out again the Hard Problem (a term with which his name will forever be associated) of explaining phenomenal experience – why is it that 'there is something it is like' to experience colours, sound, anything? The key point is that experience is simply not amenable to the kind of reductive explanation which science has applied elsewhere; we're not dealing with functions or capacities, so reduction can gain no traction. Chalmers notes – justly, I'm afraid – that many accounts which offer to explain the problem actually go on to consider one or other of the simpler problems instead (more contentiously he quotes the theories of Crick and Koch, and Bernard Baars, as examples). In this initial exposition Chalmers avoids quoting the picturesque thought experiments which are

Correspondence: Peter Hankins, http://consciousentities.com, London, UK. E-mail: peter@consciousentities.com Note: This short book reviewed appeared on my blog "Conscious Entities" at http://consciousentities.com which the editor of JCER very kindly selected to appear here.

Journal of Consciousness Exploration & Research| December 2010 | Vol. 1 | Issue 9 | pp. 156-159
Hankins, P. *The Character of Consciousness*

157

usually used, but the result is clear and readable; if you never read *The Conscious Mind* I think you could perhaps start here instead.

He is not, of course, content to leave subjective experience an insoluble mystery and offers a programme of investigation which (to drastically over-simplify) relies on some basic correspondences between the kind of awareness which is amenable to scientific investigation and the experience which isn't. Getting at consciousness this way naturally tends to tell us about the aspects which relate to awareness rather than the inner nature of consciousness itself: on that, Chalmers tentatively offers the idea that it might be a second aspect of information (in roughly the sense defined by Claude Shannon). I'm a little wary of information in this sense having a big metaphysical role – for what it's worth I believe Shannon himself didn't like his work being built on in this direction.

The next few chapters, following up on the project of investigating ineffable consciousness through its effable counterparts, deal with the much-discussed search for the neural correlates of consciousness (NCC). It's a careful and not excessively over-optimistic account. While some simple correspondences between neural activity and specific one-off experiences have long been well evidenced, I'm pessimistic myself about the possibility of NCCs in any general, useful form. I doubt whether we would get all that much out of a search for the alphabetic correlates of narrative, though we know that the alphabet is in some sense all you need, and the case of neurons and consciousness is surely no easier. Chalmers rightly suggests we need principles of interpretation: but once we've stopped talking about a decoding and are talking about an interpretation instead, mightn't the essential point have slipped through our fingers?

The next step takes us on to ontology. In Chalmers' view, the epistemic gap, the fact that knowledge about the physics does not entail knowledge of the phenomenal, is a sign that that there is a real, ontological gap, too. Materialism is not enough: phenomenal experience shows that there's more. He now gives us a fuller account of the arguments in favour of qualia, the items of phenomenal experience, being a real problem for materialism, and categorises the positions typically taken (other views are of course possible).

- *Type A Materialism* denies the epistemic gap: all this stuff about phenomenal experience is so much nonsense.
- *Type B Materialism* accepts the epistemic gap, but thinks it can be dealt with within a materialist framework.
- *Type C Materialism* sees the epistemic gap as a grave problem, but holds that in the limit, when we understand things better, we'll understand how it can be reconciled with materialism.

In the other camp we have non-materialist views.

- Type D dualism puts phenomenal experience outside the physical world, but gives it the power to influence material things,
- Type E Dualism, epiphenomenalism, also puts phenomenal experience outside the physical world, but denies that it can affect material things: it is a kind of passenger.

Finally we have the option that Chalmers appears to prefer:

- Type F monism (not labelled as a materialism, you notice, though arguably it is). This is the view that consciousness is *constituted by the intrinsic properties of physical entities:* Chalmers suggests it might be called Russellian monism.

The point, as I understand it, is that we normally only deal with the external, 'visible' aspects of physical things: perhaps phenomenal experience is what they are *intrinsically like* in themselves – inside, as it were. I like this idea, though I suspect I come at it from the opposite direction: to Chalmers, it seems to mean something like *those experiences you're having – well, they're the kind of thing that constitutes reality* whereas to me it's more *you know reality – well that's what you're actually experiencing.* Chalmers' way of looking at it has the advantage of leaving him positioned to investigate consciousness by proxy, whereas I must admit that my point of view tends to leave me with no way into the question of what intrinsic reality is and makes mysterian scepticism (which I don't like any more than Chalmers) look regrettably plausible.

Now Chalmers expounds the two-dimensional argument by which he sets considerable store. This is an argument intended to help us get from an epistemic gap to an ontological one by invoking two-dimensional semantics and more sophisticated conceptions of possibility and conceivability. It is as technical as that last sentence may have suggested. To illustrate its effects, Chalmers concentrates on the conceivability argument: this is basically the point often dramatised with zombies, namely that we can conceive of a world, or people, identical to the ones we're used to in all physical respects but completely without phenomenal experience. This shows that there is something over and above the physical account, so materialism is false. One rejoinder to this argument might be that the world is under no obligations to conform with our notions of what is conceivable; Chalmers, by distinguishing forms of conceivability and of possibility, and drawing out the relations between them, wants to say that in certain respects it is so obliged, so that either materialism is false *or* Russellian monism is true. (Lack of space – and let's be honest, brains – prevents me from giving a better account of the argument at the moment.)

Up to this point the book maintains a pretty good overall coherence, although Chalmers explicitly suggests that reading it straight through is only one approach and unlikely to be the best for most readers; from here on in it becomes more clearly an anthology of related pieces.

Chalmers gives us a new version of Mary the Colour Scientist (no constraint about the old favourites in this part of the book) in Inverted Mary. When original Mary sees a tomato for the first time she discovers that it causes the phenomenal experience of redness: when inverted Mary sees a tomato (we must assume that it is the same one, not a less ripe version) she discovers that it causes the phenomenal experience of *greenness*. This and similar arguments have the alarming implication that the ineffability of qualia, of phenomenal experience, cannot be ring-fenced: it spills over at least into the intentionality of Mary's knowledge and beliefs, and in fact evidently into a great deal of what we think, say and believe. This looks worrying, but on reflection I'm not sure it's such big news as it seems; it's inherent in the whole problem of qualia that when we both look at a tomato I have no way of being sure that what you experience – and refer to – as red is the same as the thing I'm talking about. More comfortingly Chalmers goes on to defend a certain variety of infallibility for direct phenomenal beliefs.

Further chapters provide more evidence of Chalmers' greater interest in intentionality: he reviews several forms of representationalism, the view that phenomenal experience has some intentional character (that is, it's *about* or indicates something) and defends a narrow variety. He offers us a new version of the Garden of Eden, here pressed into service as a place where our experiences are direct and perfectly veridical. Chalmers uses the notion of Edenic content as a tool to break apart the constituents of experience; in fact, he seems eventually to convince himself that Edenic content is not only possible but fundamental, possibly the basis of perceptual experience. It's an interesting idea.

Included here too is a nice piece on the metaphysics of the Matrix (the film, that is). Chalmers entertainingly (and surely rightly) argues that the proposition that we are living in a matrix, a virtual reality world, is not sceptical, but metaphysical. It's not, in fact, that we disbelieve in the world of the matrix, rather that we entertain some hypotheses about its ontological underpinnings. Even bits are things.

The book rounds things off with an attempt (co-authored with Tim Bayne) to sort out some of the issues surrounding the unity of consciousness, distinguishing access and phenomenal unity along the lines of Ned Block's distinction between access and phenomenal consciousness, and upholding the necessity of phenomenal unity at least.

It's a good, helpful book; what the content lacks in novelty it makes up in clarity. Chalmers has a persuasive style, and his expositions come across as moderate and sensible (perhaps the reduced epiphenomenalism helps a bit). It's surprising that the denial of materialism (surely the dominant view of our time) can seem so common sense.

Book Review

Stephen Hawking and Leonard Mlodinow: *The Grand Design*
New York: Bantam Books, 2010, 208 pp. ISBN: 0553805371

The Kingdom of Lies

Marc Hersch[*]

ABSTRACT

In their book, The Grand Design, Hawking and Mlodinow, faithful disciples of the scientific method, give an account of what they and their brethren in the physical sciences have discovered by following the evidence gleaned in systematic observation and measurement using the most advanced technologies available today. In their lifelong search as physicists, for the Holy Grail of a theory of everything — a Grand Design — the evidence it seems, has led them, not to a unified theory of everything, but to the heresy of all scientific heresies, a theory about theory making.

Key Words: Grand Design, Stephen Hawking, Leonard Mlodinow, scientific method, physical science, Holy Grail, theory of everything.

A Parable

In the beginning the king was told that all the crops in the kingdom would be affected by a terrible blight. Anyone who ate of them would go mad. He called in his trusted adviser and asked him what to do.

"Of course," the king said, "there is enough grain left from last year's harvest so that you and I could continue eating of it. We could remain sane and keep all the others from doing any harm."

"Your majesty," replied the wise man, "if only you and I are sane and all the rest are madmen, who is it that will be locked up in the asylum?"

"I understand," said the king, "but what is left for us to do?"

"The best we can do", replied the sage, "is for both of us to eat the same grain as everyone else but before we do I will place a mark on your forehead, and you will place one on mine, so that whenever we look at each other we will be reminded the we are also mad."

(As told to Arthur Green by Rabbi Nahman of Braqtslav)

In the human enterprise we accept that prediction is possible and it is by the device of our predictive stories — our theories — that we have prevailed as a species. In everyday experience our stories concern themselves with mundane matters of prediction: A red sky at night is a sailor's delight. The slowest cashier line in the market is always the one that I am in. The stock market went up today on news of lower housing prices. Vitamin supplements will

Correspondence: Capt. Marc Hersch, M.A, 3Sigma Systems, USA. E-mail: systems@3sigma.com Note: This Book Review is edited by JCER Editor-at-Large Gregory M. Nixon.

Journal of Consciousness Exploration & Research| December 2010 | Vol. 1 | Issue 9 | Page 160-165
Hersch, M. The Kingdom of Lies

161

make me live longer. So goes our predictive storytelling in every moment of wakeful awareness as well as in our fitful dreams.

In our never-ending quest for better prediction, we are driven to construct grander and grander stories that consolidate and reduce the number and complexity of the stories we must resort to in prediction. Given that prediction is the principal business of being human and that our survival depends on how well we do it, it is not surprising that throughout history, the human enterprise, in the grandest sense, has been to construct the ultimate story of all stories, the one true story that might confer upon us powers of perfect prediction — the story of The Grand Design.

Hawking and Mlodinow speak to the storytelling process on p. 51 of their book. A theory (an explanatory model) is more "good" to the extent that it demonstrates sensual "elegance", "is parsimonious in containing "few arbitrary or adjustable elements", is comprehensive in "explaining all observations", and "makes detailed predictions about the future" that can be tested in practical experience.

Over the ages many storytellers have laid claim to the discovery of the codex of the Grand Design — the theory that explains everything. The stars, the bones of chickens, the lay of tea leaves, or the words of gods and God miraculously revealed are but a few of the stories that have been turned to our predictive purposes, but most have fallen by the wayside, having failed one or more of the tests of "good" storytelling.

The Grand Design is a story about storytelling in which the evidence gleaned in systematic observation and measurement has led the disciples of science down a storied path of increasing elegance, parsimony, comprehensiveness, and verification in practice, to a story in which the final outcomes produced by their method of questioning may very well have brought them to a dead-end — full stop!

In the early going the authors explain that our superstitious and metaphysical mythic stories placed us at the center of the universe with all the world revolving around our being and intentions, but over time the predictive power of models that displaced us from the center and relegated us to the status of mere participants in a law-abiding world "out there" did better at meeting the tests of story goodness: "The revolutionary idea that we are but ordinary inhabitants of the universe, not special beings distinguished by existing at its center, was first championed by Aristachuc…" (p. 21).

This displacement theme, in which man's existence is subordinated to externally determined laws, forms the foundation for the world narratives of both classical science and modern institutionalized religion. In other words, this modern worldview asserts a narrative in which there is a discoverable true world "out there" that obeys the laws of nature or the laws of nature decreed by gods or God, and that by decoding these external laws, perfect prediction becomes in principle at least, possible.

In the popular press, much has been made of the idea that Hawking and Mlodinow are challenging religious thought, but the authors make it clear that this is not their aim. The authors say that they do not wish to concern themselves with the dividing line between religious stories and scientific stories, asserting that science cannot disprove the existence of God or gods. What scientific storytelling has managed to do, they say, is to tell a story in which the existence of the world we know does not "require: that there be a God or gods. A

Journal of Consciousness Exploration & Research| December 2010 | Vol. 1 | Issue 9 | Page 160-165
Hersch, M. The Kingdom of Lies

162

story is a story, and it is not the truth of a story that gives it legs, but rather its elegance, parsimony, comprehensiveness, and predictive power."

The stories told in classical science, they explain, are a product of a method of story construction: "[M]ost scientist would say a law of nature is a rule that is based upon an observed regularity and provides predictions that go beyond the immediate situations upon which it is based" (p. 27). And furthermore, "[M]ost laws of nature exist as a part of a larger, interconnected system of laws" (p. 28).

Unlike mythic and religious stories, the stories told using the scientific method of story construction must reflect self-consistent stories within stories. The authors credit Laplace with setting the gold standard of scientific storytelling, "…given the state of the universe at one time, a *complete set of laws* fully determines both the future and the past" (p. 28).

In the scientific method of storytelling, the truth-value of the lawful stories constructed can be supported by an ever-increasing number of predictions confirmed in practice, such as the rising of the Sun in the East, but a single practical falsification of a scientific story is sufficient to render that story useless, such as the day on which the Sun rises in the West. According to the authors, the first shot across the bow of the ship of truth-seeking was fired by René Descartes, who asserted the relational understanding of the principle of *initial conditions*: "In order to apply the laws of physics, one must know how a system started off, or at least its state at some definite time. (One can also use the laws to follow a system backward in time.)" (p. 20)

The authors might have better stated this in the following manner: In order to apply the rules proposed *in any story*, one must assert how the system starts off, or at least its state at some definite point in time. Every story must have a beginning, whether its initial condition be a mote in God's eye or a Big Bang.

It can be argued that the initial state for the story of scientific storytelling being told in *The Grand Design* begins with Sir Isaac Newton, a practical and God-fearing man who told a story of an interlocking mechanical universe that predicted the motions of things observed on earth and in the heavens, and a darn good story it was. Given the ability at the time of humans to observe and measure such things, his predictions were both useful and, for all intents and purposes, spot on!

Given the efficacy of his story, Newton did not have to work very hard to convince others that his story, among all others of the day, was at long last the proof of a Grand Design in the mind of God. The age of prefect prediction, it seemed, was upon us. All that remained was to employ rigorous methods of observation, measurement and testing to discover the clockwork "laws" of nature decreed by God, and thus was born the story-telling method of what the authors of *The Grand Design* call, the "classical" physical sciences. "According to the traditional conception of the universe, objects move on well-defined paths and have definite histories. We can specify their precise position at each moment in time."

The disciples of the physical sciences fashioned themselves as a monastic sect, sworn to abide by the codices of the scientific method in their quest for perfect prediction. They adopted the self-consistent and therefore perfectly true language of mathematics as their lingua franca. It was by the example of Newton's physics that all other scientific storytelling became fashioned.

Journal of Consciousness Exploration & Research| December 2010 | Vol. 1 | Issue 9 | Page 160-165
Hersch, M. The Kingdom of Lies

163

Amongst Newtonian storytellers the heretics of science came to be regarded as those who failed to heed the evidence of the senses revealed in observation and measurement. The nature of the true world, they claimed, can only be revealed if we eschew our beliefs and cleave to the empirical evidence. The history of scientific storytelling then, is the story told of following the evidence gleaned in observation and measurement conducted by ever more powerful technologies such as giant telescopes, electron microscopes, atom smashing accelerators, and brain scanners, and all was good in the quest for truth, until that is, those instruments of scientific observation began to produce evidence in which the creed of falsification itself became falsified.

Say the authors of the Grand Design: "Although [the classical science] account is successful enough for everyday purposes, it was found in the 1920s that this "classical" picture could not account for the seemingly bizarre behavior observed on the atomic and scales of existence"

So where has the latest evidence of our senses realized in natural selection and extended by technological means been leading in the search to discover the ultimate story — *the grand design*?

The evidence of our senses, enhanced and extended, about the nature of things at the smallest and largest scales of experience, seems to conspire to frustrate our best scientific storytellers, forcing them to create bizarre twists and turns of plot in order to make sense of a seemingly endless stream of self-contradictory evidence. On the scale of small, when we try to determine if light behaves as a wave or as particles, the answer depends on how we look at it. The evidence of the wave falsifies the evidence of particles and the evidence of particles falsifies the evidence of waves. The truth of the matter is as slippery as a wet eel.

When we try to determine the location and speed of a subatomic particle we find that the more we know about its location, the less we can know of its speed, and the more we know of its speed, the less we can know of its location. In the three dimensional space of our experience, one prediction precludes another.

When we shoot molecular "Buckyballs" through a slit in a screen, they pile up on the other side, honest and true, unless there are two slits, in which case the piling up is falsified, and the Buckyballs line up like soldiers in rank and file and salute us. Then again, if we peek at one of two slits while shooting the Buckyballs, they lie to us again by piling up as if there were only one slit! The Buckyballs have caught us peeking!

On the scale of the large, as we approach the speed of light, time and space are transformed in a lockstep that leads to the disappearance of both at Einstein's storied terminal velocity, the "constant" speed of light. Location in space and time along with all causes and effects are gone, baby, gone! The very foundation of our stories told in pasts and futures are obliterated.

The mathematics required to construct a story about the falsification of falsification, requires that the world we are observing not proceed along the familiar storylines of the causes and effects that mark our everyday experience in four dimensions. Randomness rules in time and space, if these places exist at all, and bounce around in 11 storied dimensions, and presumably more in some other version. Randomness rules save the evidence that some stories appear to be more probable than others and these probabilities can be practically calculated, say the authors, using the sum of all possible histories, called a Feynman sum.

ISSN: 2153-8212   Journal of Consciousness Exploration & Research   www.JCER.com
                                 Published by QuantumDream, Inc.

Now if the "probability amplitude" for one story, teased out from the mathematically calculated 10^{500} possible stories, can be singled out as greater than all the others, it would seem we are at least getting closer to fingering the story of everything we seek, but the evidence throws us still another curve ball. When we swing the bat this time, it turns out that the probabilities we calculate for a story we tell depends on what we chose to observe and how we observe it! "We create the evidence of our story by our observation rather than that story creating us" (p. 140). It does not matter which stories are actually more probable, if any can said to be so, because the story we experience as observers, however improbable, is always the one that results in us!

If the best that we can do is construct the one story, top-down, that leads to us among a multitude of possible stories that lead to universes without us, then we are returned to the center of the universe, which is precisely where we began our journey as human beings in search of perfect prediction.

The evidence, say Hawking and Mlodinow, is that there are no fixed laws of nature "out there". The world that we can observe in knowing is dependent upon the models we use, and the models we can use are determined by the conditions that lead to the one world that allows for us, among on infinitude of possible worlds.

> We form mental concepts of our home, trees, other people, the electricity that
> flows from wall sockets, atoms, molecules, and other universes. These mental
> concepts are the only reality we *can* know. There is no model-independent test of
> reality.

Hawking and Mlodinow do not doubt that there is a world "out there", but in their story they say that the overwhelming weight of evidence based in observation and measurement indicates that there is no one Grand Design within our ken. The Grand Design is "in here". Theirs is a theory about our theory-making. It is a theory about the nature and limits of the process by which we can construct stories in order to make sense of the world and predict as we go about the business of living:

> It might be that to describe the universe we have to employ different theories in
> different situations. Each theory may have its own version of reality, but according to
> model-dependent-realism, that is acceptable so long as the theories agree in their
> predictions whenever they overlap, that is, whenever they can both be applied. (p. 117)

The authors state that their best candidate for the grand design is M-Theory, in which they say the "M" stands for "master", "miracle", or "mystery", but might just as well stand for "many". M-Theory is not a single theory of everything but a theory of theory-making, in which many theories are employed to describe the universe that we observe and each story stands the test of the scientific method of story telling so long as in prediction, it does not contradict the others when their paths cross.

As with all stories, M-Theory must have initial conditions, and the authors triumphantly suggest the following:

> Because gravity shapes space and time, it allows space-time to be locally stable but
> globally unstable. On the scale of the entire universe, the positive energy of matter *can*

Journal of Consciousness Exploration & Research| December 2010 | Vol. 1 | Issue 9 | Page 160-165
Hersch, M. The Kingdom of Lies

165

(in one story) be balanced by the negative gravitational energy, and so there is no restriction on the creation of whole universes. (p. 180)

How useful is the M-Theory story that allows multiple universes, each governed by its own laws, to be spontaneously generated from lumpy randomness? What are its practical implications for the human enterprise? The authors say,

> We seem to be at a critical point in the history of science, in which we must alter our conception of goals of what makes a physical theory acceptable. It appears that the fundamental numbers, and even the form, of the apparent laws of nature are not demanded by logic or physical principle. The parameters are free to take on many values and the laws to take on any form that leads to a *self-consistent* mathematical theory, and they do take on different values and different forms in different universes. (p. 143)

The turning point in our journey in search of *the grand design* is that both the evidence and the story, as best as we can tell it, is that many if not an infinite number of stories are possible, and, given the questions we ask, some of the stories we construct will work better than others, though none can ever be perfect. The business of science, it would seem, must be transformed from the search for external truth into a search for stories that serve our purposes as creatures who make their living in prediction.

At the beginning of their book, the authors claim that their scientific storytelling has led to the end of philosophy, but in many ways their journey brings them full circle. It is more likely that the end of their story marks the beginning of another — one that attempts to unravel the question of what our purposes as predictive creatures might best be. And in the final analysis, that story can only be crafted in philosophical terms.

As with the king and his trusted advisor in the parable proffered at the beginning of this review, the evidence indicates that we are condemned to live in a kingdom of lies, in which the predictive stories we create always begin and end with us at their center. The best we can do is to place a mark upon our foreheads to remind us that we are mad and get on with the business of making our lies as useful as possible.

Journal of Consciousness Exploration & Research| December 2010 | Vol. 1 | Issue 9 | pp. 166-168   166
Smith, S. P. Review of Edmund Husserl's Book: Crisis of European Sciences and Transcendental
Phenomenology

Book Review

Review of Edmund Husserl's Book:
Crisis of European Sciences and
Transcendental Phenomenology

Stephen P. Smith[*]

ABSTRACT

All objective philosophy and positive science are unreal, that is, they all depend on pregivens that are subjective in nature. To question the pregivens is to enter phenomenology, and it is here that psychology transforms itself into Husserl's transcendental phenomenology. All "objective" science requires its purification by a transcendental psychology. Husserl (page 257) writes: "a pure psychology as positive science, a psychology which would investigate universally the human beings living in the world as real facts in the world, similarly to other positive sciences (both sciences of nature and humanistic disciplines), does not exist. There is only a transcendental psychology, which is identical with transcendental philosophy." You can find this book at Amazon http://www.amazon.com/Crisis-European-Sciences-Transcendental-Phenomenology/dp/081010458X/ref=cm_cr-mr-title .

Key Words: Edmund Husserl, crisis, European sciences, transcendental phenomenology.

Edmund Husserl's "The Crisis of European Sciences and Transcendental Phenomenology" resonates well. The following are my impressions and reflections after reading this very interesting book.

Every object-subject composite (relation) is a "phenomenon", and Husserl begins his phenomenology from Descartes' doubt that cannot be doubted. Husserl notes that the phenomenon is open to exploration. We explore so we can discover what is pregiven, so we can find our preconditions. Husserl reminds us that Kant was sterred from his slumber by Hume's skepticism. Kant's "appearance" is embedded in a space-time manifold, and as such it represents a phenomenon that hides the "thing-in-itself". The phenomenon is a composite uniting the provisional with the universal, and Kant had to feel it to be so reactive once Hume and Leibniz made their points known. Husserl reminds us to look beyond the ego-soul of Descartes, and to look beyond the dualism where Kant got stuck.

Every feeling is such a composite, so every feeling is also a phenomenon. Every feeling holds the slightest spark of awareness. I might add that every law of nature given by an equation is experiential in the sense that the law is first conceived in the mind, and then later is it empirically verified. Therefore, the law as an equation is abstraction that forgets the experiential. Because natural laws are experiential they involve feelings, and therefore these laws are phenomenological too. It is not surprising that Husserl is very critical of objective philosophy and positive science that has lost track of the subjective ingredients that come with all phenomenon.

Correspondence: Stephen P. Smith, Ph.D., Visiting Scientist, Physics Department, University Of California at Davis, CA
E-mail: hucklebird@aol.com

Journal of Consciousness Exploration & Research| December 2010 | Vol. 1 | Issue 9 | pp. 166-168
Smith, S. P. Review of Edmund Husserl's Book: Crisis of European Sciences and Transcendental
Phenomenology

167

Husserl tells us that meaning may become lost in history, and meaning relates to the preconditions of history which has to do with the geometrical horizons that history grows into. Husserl (page 49) is translated to write: "The geometry of idealities was preceded by the practical art of surveying, which knew nothing of idealities. Yet such a pregeometrical achievement was a meaning-fundament for geometry, a fundament for the great invention of idealization; the latter encompassed the invention of the ideal world of geometry, or rather the methodology of the objectifying determinations of idealities through the construction which create `mathematical existence.'"

Science grew out of traditions, and geometry is no less a tradition. The pregivens are found sleeping, Husserl tells us that the pregivens are taken for granted. Husserl (page 69) writes: "Only a radical inquiry back into subjectivity - and specifically the subjectivity which ultimately brings about all world-validity, with its content and in all its prescientific and scientific modes, and into the `what' and the `how' of the rational accomplishments - can make objective truth comprehensible and arrive at the ultimate ontic meaning of the world."

In Husserl day (right before World War II) positivist science and existential philosophy lost their meaning (I add that the meaning is still lost today), as these were all about extensions of the status quo that were no longer connected to their original preconditions.

To find the original meaning there must be a reactivation of the construction of geometry, among other exercises. Husserl tells us that meaning is discovered by reactivating the construction that have hid themselves in history. This leads us to what is self evident and beyond doubt.

The precondition of history is the stark reminder that the universal has connected with the provisional; this is the stark mystery of life, the relation again.

Husserl's phenomenology studies the precondition as it is, rather than through presumptions that derive from an extended historicism that has lost its meaning.

Husserl has much to say about intentionality, and the validation that is always sought when truth statements are attempted. And we all see people that seek validation; the pay received for a hard days work; the affirmation that is required when gifts are exchanged; the suicide note that betrays its own reason for being, as no message is needed to announce a departure unless the issue of validation is found even in the confused.

We see the need for validation in others, but can we also see it in ourselves too? Ask yourself if you seek validation in all your activities? Am I to expect an angry reaction, a denial? If so, an emotional reaction (the phenomenon again) that denies validation is an emotion that is found announcing its need for validation. In which case, the announcement is only concealed from you, but the meaning is clear to me and others that the answer is found to be yes again. If emotion is not expressed, and the answer is - yes -, then there is no disagreement. Therefore, the challenge remains to answer - no - while expressing a more reflective emotion. This challenge may be impossible to meet, as a calm denial today may follow by an angry release tomorrow, and this will cause me to return to my original conclusion: that the intentionality that seeks validation is a universal, and leads to Husserl's intersubjective person. But note also the emotional issues. It is no wonder that Husserl takes his phenomenology into psychology.

This drive to seek validity is what gives birth to our "objective" meanings, according to

Edmund Husserl, but note I put objective in quotations to refer to the observation that I am referring to a subjective transcendentalism rather than an objectivity that Husserl tells us is illusory. Science and logic can give us no help if the emotional temperament is missing, yet scientism is found today expressing its need for validation. Dawkins's "The God Delusion" is an expression that is asking religiosity to love science too. But how can religion love science if scientism lacks the emotional certitude to deal with its own pregivens? It is not unsurprising that atheist Sam Harris is now making a call for contemplation within atheistic circles. Contemplation delivers the reflective capacity to deal with our drive for validation, for both believer and nonbeliever.

Husserl (page 168) writes on elementary intentionalities that seek validity: "The being of these intentionalities themselves is nothing but one meaning-formation operating together with another, `constituting' new meaning through synthesis. And meaning is never anything but meaning in modes of validity. Intentionality is the title which stands for the only actual and genuine way of explaining, making intelligible."

All objective philosophy and positive science are unreal, that is, they all depend on pregivens that are subjective in nature. To question the pregivens is to enter phenomenology, and it is here that psychology transforms itself into Husserl's transcendental phenomenology. All "objective" science requires its purification by a transcendental psychology. Husserl (page 257) writes: "a pure psychology as positive science, a psychology which would investigate universally the human beings living in the world as real facts in the world, similarly to other positive sciences (both sciences of nature and humanistic disciplines), does not exist. There is only a transcendental psychology, which is identical with transcendental philosophy."

All of our beliefs are dependent on Husserl's pregivens, and to explore the pregivens is to enter the transcendental world that rediscovers hidden meanings of dimensionality. This activity engages our emotions, and so it is that the innate feeling is found supporting a universal grammar. As long as we remain true to our purpose, to love our self, to love others, to love God, we may always re-look at our slumber and find the hidden dimensions in our own mistakes; we can always overcome our feelings of doubt in this way, finding a deeper feeling expressed in a deeper beauty. This allows us to purify our feelings, by referring to the original intention that was never meant to do harm to ourselves, others or God. Husserl's universal drive that seeks affirmation is no more than the past that seeks wholeness with the present, it is no more than what I call the affirmation of Trinity, it is the work of the Holy Spirit among our vast plurality. This insight was meant to be shared, but in sharing this expect the emotional outcries that are found seeking their own validation.

References

Edmund Husserl, 1970, *Crisis of European Sciences and Transcendental Phenomenology*, Northwestern University Press.

Book Review

Review of John Watson's Book:
Schelling's Transcendental Idealism: A Critical Exposition

Stephen P. Smith*

ABSTRACT

Why do I review John Watson's 1882 classic, "Schelling's Transcendental Idealism"? I write this review in 2007, and the sad truth is that Schelling's system (with upgrades from Hegel, and others) is underappreciated in a world full of strife and dualistic thinking. It is underappreciated with some exceptions (e.g., Ken Wilber) even as Schelling's system could find its partial vindication coming from science. The buying public prefers its confusion coming from Richard Dawkins' "God Delusion." You can find this book at Amazon http://www.amazon.com/Schellings-Transcendental-Idealism-critical-exposition/dp/1402135688/ref=cm_cr-mr-title .

Key Words: Schelling's system, transcendental idealism.

John Watson does a wonderful job describing Kant and his aftermath, describing Fichte's thinking before moving on to Schelling. Watson follows the movements up to Hegel's entrance, but Watson writes mostly about Schelling's contributions.

Watson (page 98) writes: "Even more strongly than Fichte, Schelling rejects as absurd and unthinkable any `objective' God, independent of man and nature, and seeks to explain each entirely from itself. " Schelling's God could not be held separate from God's creation.

Watson establishes "the fundamental proposition of philosophy", and writes (page 109-110): It is not only the supreme condition of knowledge, but of action as well. Assuming, in the meantime, that a knowledge of objects is possible, and that volition also is possible, it is evident that both alike presuppose our fundamental principle. There can be no knowledge of anything apart from consciousness, and, as has been shown, no consciousness apart from the self-activity which we call self-consciousness; nor can there be any volition which is not in consciousness, and therefore none which is not made possible, and alone made possible, by self-consciousness." Schelling, we are told, develops his transcendental philosophy beyond Kant by recognizing two acts of intelligence: pure activity as volition and the limit of that activity presented as sensation.

Watson (page 117) writes: "Sensation is not a mere limitation, but a consciousness of limitation, and such consciousness necessarily presupposes that there is, at the very least, a reaction of consciousness against that which is opposed to it."

Watson (page 122) writes: "Perception is not the purely subjective apprehension of an independent object, but the actual apprehension of an object existing in relation to consciousness."

Watson (page 180) writes: "The world is a divine poem, and history a drama in which

Correspondence: Stephen P. Smith, Ph.D., Visiting Scientist, Physics Department, University Of California at Davis, CA
E-mail: hucklebird@aol.com

individuals are not merely actors but authors; but it is one spirit which informs all and directs the confused play of individuality to a rational development."

Unity in opposition is simple enough in principle, yet sometimes a sensation comes from reading Watson that a better articulation is possible (either from Watson or Schelling). This unfinished quality is apparently the nature of the beast; next to the infinite us finite folks are somewhat incomplete. It is better to admit our incompleteness and this is to discover our best art, even in the handiwork of an artist yet to be. Watson (page 194) writes: "Perhaps it is not unfair to say that no amount of self-restraint could ever have enabled Schelling, with his quick imaginative temperament, to build up such an edifice of philosophy as his great successor Hegel has left to us." It is worth noting that Hegel is not well understood today, but perhaps that will change.

References

John Watson, 2005, *Review of John Watson's Book: Schelling's Transcendental Idealism: A Critical Exposition*, Adamant Media Corporation.

Made in the USA
Middletown, DE
06 November 2019